THE ELECTRONIC
GIANT

THE ELECTRONIC
GIANT

A Critique of the Telecommunications Revolution from a Christian Perspective

Stewart M. Hoover

The Brethren Press, Elgin, Illinois

Library of Congress Cataloging in Publication Data
Hoover, Stewart M.
    The electronic giant.
    Includes index.
    1. Television broadcasting—Moral and religious aspects. 2.
Christian ethics—Church of the Brethren authors. I. Title.
PN1992.6.H6   261   81-6083
ISBN 0-87178-217-0   AACR2

To Mom and Dad, who worked
so hard for me

Contents

Foreword

My mother gave me sound advice a long time ago. "A responsible parent (friend, lover, teacher) is a person who will warn you about poisonous snakes if you are about to cross a field filled with them." This was not a suggestion to prevent the adventure of exploring new regions or a prediction of failure. She was simply convinced that those you love deserve to know the lay of the land. Stewart Hoover has fulfilled my mother's injunction with this fine book.

We are thrust into the midst of an environment radically different from the past which was the basis of our education, socialization, faith formation and family upbringing. The heritage we offer to the future has been cast in pre-telecommunication molds. This pinch between old and new forms creates a crisis for many who bear the past to the future through the present. For persons crossing this stretch of cultural wilderness, life can be a frightening and confusing experience.

Stewart provides a terse introduction to the delightful and dangerous media environment lurking before us. These serpents of challenge come to us via wires, lasers, waves in the air and beams from the satellites. We can't have our teeth drilled, ride an elevator, cruise a freeway, visit a funeral home or use a public restroom without the influx of music or other material into our consciousness. The author offers an intensive look into the unfolding telecommunications world of the present and future.

I particularly appreciate his ability to steer away from the extreme viewpoints which have seduced many commentators on mass media in recent writing. They either paint a picture of hopelessness for those who pursue theological values in the midst of the media revolution, or they describe a wondrous new world where the home computer and other telecommunications opportunities will usher in a future free of past problems. Stewart happily leads us on a trek into the future present with a clear eye. He depicts both the hopes and the dangers of the electronic society. In other words, this book provides information about the unknown meadow before us. As a good friend (lover, teacher), he knows that you deserve to know about the dangerous coiled creatures in your path. Yet, he invites you to take the journey into this adventure. May the trek be exciting for you while you avoid the fangs of despair.

Dennis C. Benson

Preface

I spent the first years of my life in a small western Nebraska town that did not have even paved streets. Our telephone hung on the wall and had to be cranked to raise the operator. We had radio, and I can remember listening to episodes of "Fibber McGee and Molly."

My father made a media critic of me at an early age, however, by subscribing to *The Sunday New York Times* by mail. My parents' intellectual tastes and needs were not satisfied by the small-town papers available in the area. Thus our family probably was unique in the community because some of our cultural reference was to New York City, a place most of our neighbors had never visited and probably never would.

Our unique perspective diminished, however, when a television translator station was constructed nearby. It brought in a television station from across the state, and suddenly New York City was there in our living room (and everyone else's *before* ours—we were slow to get TV), but in a form different from that brought by *The Times*. For the first time, I saw a "live" version of New York, the gospel according to "The Edge of Night." I also vividly remember that serial presenting the first shooting I saw on television.

Other changes have come to that small town in subsequent years. There are probably dial phones now, and more than one television station. Soon, there will be a plethora of new communications technologies available to that sector of the world once parochial and isolated, as well as to every other community in the country.

What we were witnessing there, in Enders, Nebraska, in 1954, was the emergence of a new system of communications—television—that was qualitatively different from the ones that had gone before it. In Enders we skipped from the age of crank telephones to the age of TV, a faster evolution than was experienced by people in larger cities, but much more gradual than the evolution in other small towns in Nebraska and Colorado that had television even before they had telephones.

We were probably not aware of how these events would shape our lives in the years to come, although some effects in our little household were quite visible. Instead of doing things as a family in the evenings, we sometimes watched TV. Our subscription to *The New York Times* eventually lapsed, for we could get national news over television (though Dad was never quite satisfied with the trade). We were certainly not aware of the broader societal ramifications of television, though anyone who lived through its advent can probably put together some sense of those broader issues.

We are in the midst of a revolution—a revolution in the way we communicate as a society. As with past revolutions we will be and are being changed by it. This revolution began with the coming of the medium of television and its extensive penetration into our homes. Its effects can be as small and subtle as my experience in Enders, but they will also be felt at all levels of our society and culture.

A revolution of similar scale occurred with Gutenberg's introduction of movable type, but we have been conditioned by high school history to look on the effect of Gutenberg's invention as momentous and instantaneous. We thus expect that if anything so grand were occurring today, we surely would notice. In fact, the full effect of printing took years to work itself out. Indeed, its effect still ripples through contemporary institutional life in such forms as controversy over the importance of literacy training in national development. Most of the people alive at the time of Gutenberg probably were not touched by his invention during their lifetimes or the lifetimes of their children.

Contemporary scholar Elizabeth Eisenstein has cast a new light on Gutenberg in a recent book, *The Printing Press as an Agent of Change*.[1] Eisenstein suggests it was not only the manufacture and distribution of books that had an effect on society, but also the manufacture and distribution of printing presses. These presses established a new class of intelligentsia, the publishers, around whom cells of writers and thinkers gathered in hundreds of decentralized locations.

A superficial analysis of Gutenberg might have looked at the immediate effects of hundreds of Bibles being distributed (which would have been minimal unless literacy rates also allowed them to be read) while a deeper, longer-lasting social effect was related to the way Bibles were produced.

So it is with modern mass media. The institutional aspects of their production and distribution—the organizations involved, their operations and policies—are at least as important for evaluation as their content or their "effects."

The institutions and systems of the mass media work together as an integral component of our society and culture. They work to set the agenda of what is and what is not appropriate for us to consider. We tend to allow these institutions to act in this way, powerfully yet subtly shaping what we know and how we react.

As we enter the new "telecommunications age," it is time for us to reconsider how we relate to these powerful forces in our lives. We need to begin to declare our autonomy by turning our attention to the very institutions through which we get most of our information about the world. How are they affecting us? How should we respond? How can they be changed?

To answer these questions, we must begin with a basic knowledge of these institutions, of the technologies that are fueling the "new age" and how these things are converging to transform our daily lives. Such knowledge should be considered adult basic education. Just as we need to have a basic understanding of the legal system or banking system in order to survive in today's world, so should we have an understanding of the network of institutions that have been called the "fifth estate"—the mass electronic media.

Maybe more important to those who would wish to evaluate these trends from a Christian perspective is knowing how these developments affect ministry and witness in contemporary and future society. The telecommunications age will transform (and is transforming) the field on which social issues can be addressed. It may pose theological challenges as well. Now is the time for Christians to begin developing skills and awareness in this area. We need to assert our right and responsibility to a perspective on these developments.

This book is intended to help form this Christian perspective. It can be used by students, teachers, pastors, parents, media reformers and others interested in the mass media. Little attention has been given to media ethics and processes from the perspective of the church. As one societal institution that is free to take moral and ethical stands, the church can find an important mission in this endeavor. Other institutions, including the schools, need to begin looking at these developments as well. The purpose should not be to oppose them without knowledge, but to honestly evaluate them, to separate the good from the bad, and to develop strategies in accordance with those evaluations.

The time is urgent, the challenges are real and exciting. We must work to become "new people for the new age"—people who are radically aware of its challenges, and prepared to meet them in our personal and corporate lives.

I wish to acknowledge the teachers who have helped me during my personal quest for answers to the basic questions presented here. Among them I would mention: Dale Goldsmith, who taught me the Bible; Dennis Benson, who introduced me to the media; Karen Lebacqz, who was my mentor during my period of ethical inquiry; George Conklin, who helped me understand the nature of the beast; Howard Royer, who let me take action; and George Gerbner, Robert Shayon, and Barry Cole, who have given me helpful guidance more recently. I would also acknowledge the diligent work of my editors, Harriet Ziegler and Fred Swartz, who have fought tirelessly to save the project from rack and ruin at my hand. Finally, I would especially like to acknowledge the love, care and assistance of my wife, Karen, who stood by me while I worked on the book, but who wanted to see it only after it was completed! Whatever value there is in this volume has been made possible by the help and care given me by these people.

Stewart M. Hoover
Philadelphia, Pennsylvania
April 14, 1981

Section I

Ethical Implications

Introduction

Much of what has been said about television in recent years has concerned its relationship to the ethics of its viewers, and its ethical implications as an institution. Because the term ethics has been used to cover such different issues, there has been much confusion about whether television has an ethical dimension and, if it does, what that dimension entails.

Television most certainly does have ethical implications. These implications concern individual viewers whose ideas about reality and normal behavior may be affected by their TV consumption. These implications also concern the public as a whole and the prominent institutions of society, because television has changed some of the basic processes and relationships of society and culture.

In order to honestly evaluate these things, the pitfall must be avoided of looking at too narrow a slice of the phenomenon of television. The available evidence about television's role also must be considered.

Much of the data on television appears to be contradictory. One study claims that television has an effect on children's behavior, another questions that finding. It is the task of the intelligent observer to arrive at a synthesis of such data. The apparent contradictions are due to the complexity of television's interaction with society. There is enough diversity in that arena that evidence can be found to contradict nearly every research finding about television.

A consistent thread to most of the research, however, suggests that television content definitely affects certain viewers in certain ways. The fact that not all viewers are affected similarly should not detract from the direction of the evidence. Those who want to assess television's effect must arrive at a holistic perspective. Although it is necessary to examine all forms of research on television's effects, it is imperative to move beyond a survey of research so that response is not drawn from too narrow a perspective.

Chapter 1

The Medium, the Message and More

The telecommunications revolution may have begun at the 1939 New York World's Fair where the Radio Corporation of America displayed the first production model television sets seen by the public. They were huge contraptions with picture tubes that were so long they were mounted vertically with the screen at the top. A set of mirrors redirected the visual image out the front where it was viewed through a lens. The image was poor by today's standards but reportedly captivated those who saw it.

Twenty years later, this new invention had supplanted the radio as the dominant medium in the American home. But it was an experience different from radio in kind as well as degree. The glowing screen brought visual as well as auditory stimulation to the audience and, many people suspected, placed a lower demand on the audience's imagination and creativity.

Aside from data on audience sizes there is little scientific evidence about the difference between the way people used radio before TV and the way people use television. They are different media, however, and even today radio use is high, partly because it can be listened to while the listener is engaged in other activities, such as driving, where watching television would be impractical.

As had been the case with earlier media—films, penny novels, and radio—great public concern arose over television's effects on its audience, primarily its effect on ethics and values. Were viewers and listeners being "corrupted" by this experience? Were they learning lessons that would affect their behavior?

Generally, ethics is a term used to describe the act of considering questions of what is right, what ought to be done, and what the responsibilities are. Thus, ethics can refer to anything from simple, circumscribed codes of rules, such as professional ethics, to consideration of the complex issues of right and wrong involved in euthanasia or bioengineering.

The issues in telecommunications ethics go far beyond the industry code and encompass the complex picture. A professional "ethic" covering activities of a specific broadcasting organization may describe a code consensual among those broadcasters, but the *audience* to whom those broadcasters relate is the object of, not a participant in, that process, and thus a higher order of concern is implied. Professional ethical codes are not bad things, but they are often designed for far different purposes from those to which the term "ethics" might be applied. Rather than arriving at what is right in an ultimate sense, professional "ethics" may encompass only what is right within the law or only what is needed to avoid the behavioral enforcement from outside sources. Such has been the history of some sections of the broadcast code of the National Association of Broadcasters (NAB). Barry Cole and Mal Oettinger, in their history of the Federal Communications Commission, indicate that much of the NAB code is there because of threatened regulation by the FCC.

In his book, *Television: Ethics for Hire,* Robert Alley provides a comprehensive and thoughtful description of television's production and its motivations and intentions. Alley's perspective is summed up in his description of television in the book's introduction:

> Television may have many practical uses, but its primary character lies in two areas: as a conveyor of information and as an artistic medium, a maker as well as a reflector of cultures.[2]

His analysis thus stresses the *content* of the medium—specifically the intention and perspective of those involved in production of the content. His report is fascinating and successfully counters what he sees as the primary shortcoming of television's critics over the years—a sort of naive distaste for a medium that one of them described as "bubble gum for the eyes."

A consistent thread of antielectronic media thought definitely exists among the intelligentsia of America. The electronic media have been taken much more seriously in other Western countries. This lack of concern for electronic media has been as true in the church as anywhere else. Television has been dismissed as either irrelevant to theology and faith, or as the concern of those on the fringe who decry its "sex and violence."

Alley argues that an ethical analysis of television reveals it to be an important source of information about ethics and values, worthy of more serious concern and consideration by the disciplines of theology, philosophy, history, ethics, and literature. He falls short, however, of describing television's other ethical implications in a comprehensive way. His finding that there are ethical and moral intentions beyond program content is supportive of the general idea that television is important in these regards, but he tends to dismiss criticism of television on moral grounds as either naive or ill-advised.

Such a perspective is too narrow. The complexity of the audience's interaction with television has made it difficult for social scientists to determine exactly what the effects of its contents are. Thus the assumption by producers that their effects actually work out for the audience as intended may be false. Even beyond that, it must be remembered that television is a larger, more complex process than merely a production of art, entertainment, or information. How does the audience react? Where will the program be placed? Who will see it? What advertisers will buy it? What will happen in its censorship by the stations and the networks? Is the "idea" intended also carried out? All of these questions bear on television, and the answers to any of them give important clues to the moral, ethical, or values impact of a specific television experience.

It is important to see television and the other mass media as social issues of the day, but also they have become issues themselves and need to be taken seriously. Succeeding chapters will discuss the ways the media have come to be important issues, and look at them with an eye to their moral and values implications.

The organization that awards the "Emmy" to outstanding television programming has an ostentatious title: the National Academy of Television Arts and Sciences. Contrary to that title, television should be seen as neither an art nor a science, but as a *craft*. Rather than carrying either the hard, consistent discipline of a science or the independent freedom of an art, television combines both of those characteristics with a continuous attention to production for economic gain, like an artist who might turn to the production of large numbers of craft items for public sale. The emphasis on economic participation changes the focus of the artist's work.

The primary ethical judgment that might be made about television and the other mass media is that it is not enough to look only at any one aspect of their institutional processes. Each person's evaluation of television and the other mass media can and should take into consideration as much information as possible about these institutions. This may seem to be a difficult task. The weight of information to sift through makes it seem that "the more we know, the less we know." Observers should be prepared, however, to hold the various uncertainties in tension, and move at least to some action based on what they discern. This book intends to provide the sort of information about television and the mass media that will aid the reader in intelligent consideration of these ethical, moral, and values implications.

Television may interact with ethics and values in three specific ways: it could provide information useful to viewers in making ethical or value judgments; it could provide reinforcement for already held beliefs about value and behavior; it could stimulate certain kinds of behavior through direct modeling. All of these "effects" of television have been found to occur in some degree. A separate area of concern, the institutional issues that have been stressed here, will examine the way the earlier electronic media—particularly television—developed, and then move to a discussion of the "new media" of the "information age" and speculation about their probable development and effect.

Chapter 2

The Dubious Authority

Television as an issue is difficult to grasp because it is both an institution of society and a pervasive experience in which all share. Because it is a pervasive experience, part of television's role is to inform the viewer, part is to train the viewer how to be informed. Because it is both an institution and an experience, the distinction is made less clear between an institution from which certain information comes and an institution that creates a framework for evaluating that information. This dual nature makes television a dubious authority.

Much of the speculation of communications theorist Marshall McLuhan is difficult to translate into practical terms but he was right in describing the new media as qualitatively different from earlier ones. They are consumed and perceived differently. Beyond issues of perception, however, communication is by its very nature an ordering process, redefining the relationships between those involved whenever the media change.

Telecommunications

Telecommunications is a term that primarily refers to electronic communication and traditionally has been related to such things as telephone or cable communication rather than to broadcasting (television and radio). Due to massive changes underway in the communication system in the United States, however, the distinctions among television, radio, and more traditional telecommunications services are becoming blurred. Television in particular, and to a lesser extent radio, are being drawn into a larger telecommunications system, a process that will take a number of years to complete. When it does happen, however, a fundamental shift will have occurred in the way society is organized socially, economically, and intellectually.

Clues to the way societies will be changed by this process are available through study of the past. The "home information center" of the future will be similar to the television set of the past in more ways than appearance. Lessons learned in the television age should help in discerning the demands and changes of the telecommunications age. The communications developments of the future have been set in motion by the communications experiences of the past partly because communication *is* order and system. The system and order by which people have been socialized in the past should help predict their preparation for certain kinds of systems in the future.

Mass Communication

Sociologist Charles Wright has outlined four ways mass communication has been conceptualized. According to Wright, some see it as a "technologically magnified form of private personal communication," such as the oft-noted observation by viewers and performers that somehow television brings them into contact with one another.[1] Another school of thought holds that mass communication can be seen as a rather mechanical "input-channel-output" system such as Shannon and Weaver's transportation theory diagramed below. This "mechanical" approach, according to Wright, also includes analyzing mass communications as an "information system" where centralized sources send messages to many receivers.

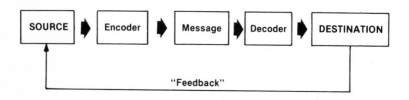

Basic Shannon-Weaver theory of communication (adapted from C. Shannon & W. Weaver, *The Mathematical Theory of Communication* (Urbana: University of Illinois Press, 1949)

A third way of approaching mass communication, according to Wright, has been the "technological determinism" of Marshall McLuhan, Harold Innis, and James Carey. This perspective holds that societies are transformed by communications technology, almost irrespective of its content.

Wright's own perspective considers the sociological ramifications and functions of mass communication, noting the major activities of mass communication to be (1) surveillance of the environment, (2) correlation of the parts of society in responding to their environment, (3) transmission of the social heritage from one generation to the next, and (4) entertainment.[2]

These functions of the mass media, first elucidated by pioneering researcher Harold Lasswell, describe activities that at one time were the province of other social institutions, primarily school and church. As the use of television has grown with respect to other institutions, its importance as a conveyor of social heritage and cultural information also has increased. Television may not have supplanted the school and the church completely, but it and other mass media have certainly encroached on the traditional turf of those institutions. Teachers find themselves either fighting television's powerful influence or cleverly using television as a tool, an aid, a discussion starter. Ministers have found more and more illustrations from television creeping into their sermons. In religious education classes, television provides a kind of common experience which, while nearly universal, also limits and "canalizes" the range of issues that can be discussed.[3]

Concentrating only on the effects of mass communication in terms of what its content "does to" its audience is narrow. This view ignores the fundamental reality of the *craft* of mass communication—that it is an institutionalized production process. George Gerbner, a prominent communication theorist, has suggested that the mass production of messages may be a more important feature of mass communication than the technological feat of linking a central source more or less simultaneously to a number of receivers.[4]

As society moves from the age of broadcasting to the age of telecommunications, Gerbner's perspective becomes even more important. To a great degree, the new media will individualize the messages sent from a source to each audience although the institutional production of the messages will continue.

In the broadcasting era, much can be learned about the experience of mass communication by evaluating the content of a few centralized transmitters, such as the three networks in the case of television. Most people watch network television most of the time, and some conclusions can be drawn about the fact that their experience is more or less universal in terms of time and content.

A common tactic among those who want to sidetrack public criticism of television is to say, "All of this concern will be moot in a few years. The networks as we have known them will go out of existence with the coming of cable. Then we'll be in a whole new ballgame." This thinking reflects the belief that public concern about the quality of broadcasting has been based on the structure of the medium: a centralized source disseminating messages to a mass audience. It is true that the structure will change in the new age, with fewer and fewer people and more and more sources (though the proliferation of sources may prove illusory). Other important characteristics of the process, including the way it attracts its capital and, more important, its role as an institutionalized mass production process, will continue to hold true.

Therefore mass communication must be considered both from the perspective of its structure under broadcasting and from the perspective of its institutional processes. Its processes provide a bridge to the "new age" of telecommunications when, while the game will appear to change, the rules will remain the same. Limiting inquiry to the content of mass messages while ignoring the larger issues of structure and process would be analogous to limiting inquiry into the policies of an infant formula distributor to the quality and content of the formula, ignoring the process by which it is made and marketed.

The "effects" of television

In a Television Awareness Training workshop participants were asked to write a concern they had about television. One of the responses was most poignant. It said simply, "It concerns me when my three-year-old's cat dies and she says the policeman shot it." This was in a town too small to have any police other than a part-time constable. Certainly the only way that idea could have come to such a young child was through the medium of television which brought the "big city" police into that little town daily.

When this story was recounted to a group of educators later in a larger city, one of the listeners, a sociologist at a local university, reported on a project for young children (aged six to ten) who were taken on tours of the police academy in order to promote understanding between inner city kids and police. In an early "intake" questionnaire the kids were asked, "What do police do?" She reported her surprise at finding a large number answering, "They shoot people." In the not-too-distant past, the answer expected would have been, "They help lost kids" or "They give directions" or some such. Admittedly, inner city children are more likely than others to actually have seen police shoot someone, but it is a relatively rare experience even for them. A much more likely answer based on inner city experience would be something like, "They ride around in cars."

The Washington Post of July 29, 1976, reported on the shooting death of a six-year-old boy by a three-year-old and his six-year-old brother. According to the report, the loaded gun was left in their home by their father, a security guard. They had an argument with the victim and went home to get the gun. The six-year-old cocked it and the three-year-old fired it, killing the other boy instantly. When apprehended by neighbors and police, the boys were home watching television. While most of the blame for this must lie with anyone who would leave a loaded handgun around the house, the fact that the boys knew how to cock, aim, and fire the gun, and more important, saw the use of a gun as an appropriate solution to conflict, implicates television in this situation.

In a university class where the students were asked to write papers on whether violence on television should be banned, one woman argued vehemently that it should not because television is entertainment, irrelevant to violent behavior. In support of her case, she recounted an incident that had occurred while she was working at a day care center. She had found a group of preschool children throwing blocks at one another from behind forts made of chairs. She stopped them and asked why they were behaving in such a way. They were surprised and dismayed at her anger because they were "only playing Star Trek" and the blocks were their "ray guns." The student argued that the incident was really trivial, that the involvement of television was minimal, only a stimulator of fantasy. In reality, this is an excellent illustration of the concern about television violence. The "fantasy" became quite real to children who were hit on the head by blocks regardless of whether they were "ray guns" or not.

The phenomenon of air piracy began in 1967 when an Australian jet was hijacked by a man with a gun. Almost overlooked initially was the fact that the incident was a copy of one on a television drama that had been shown a few days earlier on Australian television. A rash of copycat incidents has now led to security checks at all airports, a major inconvenience to passengers.[5]

One media reform activist has said, "Now that's an effect of television I feel everytime I fly." He does not necessarily mean that such a situation would not have occurred without television, but he realizes that the existence of television has affected everyone.

A more recent chapter in the air piracy story was written in 1980 when six aircraft were hijacked to Cuba in eight days. All of the hijackers used the same technique, smuggling flammable liquids on board in plastic containers. The hijackings were widely reported by the media, including the technique used in the early ones. It is probable that at least some of the later hijackings were suggested by the dissemination of that information.

Two other incidents have received wide public attention—the 1978 "torching" of a derelict by a group of teenagers in Boston who claimed they got the idea from a television show the night before, and the trial that same year of Ronnie Zamora, a Florida youth who argued that television was to blame for a murder with which he was charged. The Zamora case received wide press attention. He was found guilty and the television defense thrown out. Although television could have been implicated as a stimulator or modeler of the behavior Zamora carried out, society simply could not create a precedent for passing off the responsibility for such acts to television. This fact, however, should not lessen concern over television's role in this and other antisocial behavior. Individuals must be responsible for their own actions, but society should be prepared to lessen the pressures on those individuals—be they economic or emotional— which lead toward antisocial behavior.

The above incidents illustrate two different areas of concern about television's effects. The young girl whose cat died and the inner city kids responded to television as a conveyor of information about the world, information that could lead to certain kinds of behavior. The boys with the gun, the children playing with blocks, and the others seem to respond on a more concrete level, actually imitating the behavior they observed. More children are affected in the former way than the latter, but both types of response are explained by the scientific theory of *observational learning,* the basis of most television violence research.

Robert M. Leibert, a prominent television researcher, has described a process of observational learning involving three states. First, the child/viewer must be exposed to a behavior; *observation* of a behavior must take place. Second, the behavior may be *acquired;* the child remembers the behavior and the context in which it may be appropriate, integrating it into his or her bank of behaviorial knowledge. Third, the behavior may be *replicated* (repeated) by the child in similar or dissimilar circumstances.[6]

Research on television's effects on behavior began in 1960 with a landmark study by Albert Bandura to determine if television had any relationship to this observational learning process. Before Bandura, there was some doubt about whether viewing film or television would have any relationship to behavior (in spite of the fact that advertisers always had seemed convinced).

Bandura exposed a group of youngsters to a film modeling aggression and placed them, along with other children not so exposed, in a lab setting resembling the film situation. A far greater number of those who saw the film repeated the violent behavior than did the others, thus supporting the theory that television could be active in teaching aggression.

Additional studies were done throughout the 1960s, but the whole research area received increased attention beginning in 1968 when the Presidential Commission on the Causes and Prevention of Violence raised questions about the possible involvement of television in this area.[7] The surgeon general of the United States commissioned a study of the specific issue of television violence in 1969. The report was released in 1972. This controversial commission was the target of intense lobbying by the broadcast industry which resulted in some researchers, such as Bandura, being blackballed and other, more "pro-industry," researchers being substituted. Leibert reports that the persistence of these industry representatives resulted in a plethora of qualifying language inserted in the final commission report, adding to the perception that the findings were less than definitive.

> Certainly my interpretation is that there is a causative relationship between televised violence and subsequent antisocial behavior, and that the evidence is strong enough that it requires some action on the part of responsible authorities, the TV industry, the government, the citizens.[8]

The surgeon general's studies showed a definite role for television in teaching violent behavior to children, by bringing a wide variety of violent incidents into the home for children to observe (established by studies called "content analyses"), by stimulating direct imitation of violent behavior (in *"laboratory studies"*), and by teaching other lessons about kinds of aggression and when they are appropriate in real life situations (in *"field correlational studies"*). A variety of research strategies has been chosen for these studies and more recent ones because television cannot be studied in the more traditional scientific way in which a "control" group without television is compared to a group with television. There are not control groups available in the United States since virtually everyone has television. Therefore, research must use an approach that attempts to determine if television does carry violent messages (it does), if children learn and remember violent messages they see (they do), and, most important, if they repeat those violent acts in play or other situations (they do).

More recent research has supported the surgeon general's findings and has extended the study to older children and to adults as well. An interesting study was funded by CBS at a university in England in 1978. This study, surely designed to give the maximum benefit of the doubt to television, still found a significant increase in violent behavior and antisocial attitudes among adolescent boys. These effects were also relatively large and long-term. This study confirmed one of the most striking of the earlier surgeon general's findings: For a group of young men followed over a number of years in a longitudinal study, their level of personal aggression at age 19 was found to be most significantly related to their exposure to television violence ten years earlier, at age 9, regardless of other factors such as parental income or education level.

Clearly, not every child is affected by television all the time and in the same way. Even according to the theory of observational learning, there is a variety of reasons why violence on television may not lead to violent behavior. The child may not see the behavior depicted or may be distracted. The child may not remember it after seeing it or may not remember it in such a way that it can be recalled or repeated. A multitude of other factors in the environment may inhibit the child's memory of the behavior or inhibit the repetition of it.

One of the most definitive findings of the research has been the discrediting of the *catharsis hypothesis*. This hypothesis suggested that rather than stimulating aggression, film or television would dampen it by providing the viewer a vicarious experience that allowed release without resorting to real violence. With little support now for this theory, the reality remains that television can and does teach violent behavior, and that when that behavior does not occur it is because of other factors, *not* because of a character of neutrality or antiviolence on the part of the television experience. Television seems to have the overall effect of increasing the likelihood that children will know how to perform violent acts, will have specific ideas of when those acts are appropriate, will feel that violence is a normal behavior in society, and actually will be more violent or aggressive in both actions and language.

Television may have effects beyond the behaviors studied most frequently. A number of the above examples dealt with an *informational* effect of television that did not necessarily imply imitative behavior. The children's beliefs about the role of police in society did not carry a behavioral component like the boys' use of the gun did, but the fact that such information is being widely learned from exposure to television implies the values-related effect of this medium.

Another effect of television might be the viewing of the medium itself. In recent books, the thesis that *viewing* is the important factor, not content, has been extensively discussed. Theories such as those which suggest that television may interact with one side of the brain, may be a "hypnotic," unconscious experience, or may be a sort of narcotic, have all been put forward. There is not much data on these ideas yet, but it is at least the case that life today in the average home is different because of the presence of television.

Mealtimes have been standardized to coincide with or avoid the evening news. Television babysits the children. Children learn that adults use television as an escape, and repeat that behavior themselves. Television inhibits evening visiting, a major pastime in former years. Pastors and other professionals find home visitations affected by the presence of television. Scores of "presence of television effects" exist and their importance will increase in the new age.

Another values-oriented effect of television might be in a less-researched field, the field of *moral reasoning,* or possible effects of television on the *way* moral or ethical dilemmas are decided. The whole idea that there was such a thing as moral reasoning and, further, that it could be taught, fell out of disfavor after early studies of the effects of self-conscious moral education by Hartshorne and May, Havinghurst and Taba, and more recently, Stanley Milgram.

Hartshorne and May's study was illustrative of this area of inquiry. They sought to confirm the value of a "moral" oriented education by studying the difference in response to a tempting opportunity to cheat. One group studied consisted of children in conventional school settings; the other, of children who attended parochial school and/or were members of organizations such as the Boy Scouts. Their finding was that the probability of being caught was the thing most related to the likelihood to cheat, and that the response of both the "moral" and "nonmoral" educated children was almost identical. The "moral" kids were even slightly more likely to cheat in some cases.

This and later studies led to the widespread belief that morality is not something that can be "taught." This conclusion, coupled with the reluctance of public education to involve itself in moral stands, has resulted in a paucity of attention to the process of making moral decisions. When it was discovered that teaching children values as rules (rather than as ways of reasoning) did not necessarily result in their acting "virtuously," the whole enterprise was rather summarily abandoned.

It has been only recently that new moral education efforts have been undertaken, particularly by disciples of Jean Piaget in Switzerland and Lawrence Kohlberg in the United States. Piaget found that young children did indeed reason morally, and that they went through a developmental process in which they reasoned differently as they matured.

Kohlberg extended Piaget's theories to adolescence and later, finding that a developmental sequence in reasoning extends into adulthood. Some of Kohlberg's speculation about Piaget's theory has come under criticism recently, but his scheme of stages and the relationships between them provides a useful tool for evaluating moral reasoning as it occurs, and for looking at the quality of moral statements that are made in a medium such as television. Kohlberg's stages of moral development are:

1. *Punishment and obedience orientation.* The young child responds to dilemmas of should and ought entirely on the basis of avoidance of punishment. Identification with an external authority (parent or teacher) is greatest at this stage.

2. *Instrumental relativist* or "reward orientation." The child begins to act consistently in order to achieve positive, as well as negative, reinforcement.

3. *Interpersonal concordance* or "good boy/nice girl" orientation. The identification shifts somewhat from the dominance of an external authority to the desire to be liked by all external authorities, including peers. Moral judgments are based on the child's desire to be liked.

4. *Law and order* orientation. The individual sees the important factor in any moral decision to be the existence of a system of rules by which all people must live.

5. *Social contract* orientation. The individual's own moral judgment comes to dominate and he or she recognizes the fact that the rules governing behavior are the result of social consensus.

6. *Universal ethical principle* orientation. The universality of certain principles predominates for the individual and he or she comes to a certain independence of decision making.

It is easy to see how Kohlberg's system could help explain some of the findings of the earlier studies, particularly the cheating study of Hartshorne and May. Children at stages 1 or 2 in Kohlberg's system would certainly appear to have no moral capacity whatsoever in Hartshorne and May's situation. Using Kohlberg's system, however, it could be concluded that they were merely at a lower stage of development.

A number of other points are useful in understanding Kohlberg's theory. Most adults in America reason primarily at level 4. Regardless of one's level of development, it is always possible to reason at one of the lower stages. People who are attracted to the thinking of a stage above theirs find it superior to their own if asked. People move up in stages when they are confronted with situations where their own level of development is inadequate preparation for them to reason.

It is unlikely that television could be active in stimulating shifts in moral reasoning, particularly in the upper stages. One study, however, has found some reason to suspect that television does present moral reasoning at specific stages (never over stage 4) so that there could be an impact of moral development along Kohlberg's lines.[9]

This impact could be of four kinds: (1) For Kohlberg and Piaget, specific "real world" experience of dilemmas leads to growth. Television is heavily viewed by children at the expense of such real experience. (2) Television's presentation of moral dilemmas may provide experiences for thought and discussion by viewers, another important process leading to growth. (3) Television's orientation to the presentation of moral reasoning of a stage above the viewer's could be attractive to the viewer. (4) To the extent that television's reasoning is primarily stage 3 (think of commercials!), any stimulation of reasoning above that level would not occur.[10]

Television and Adults

By the time people reach adulthood, their behavioral patterns are well established. Unless adults put themselves through major resocializing experiences such as changing jobs, going to college, or joining the military, the chances of their behavior being influenced by television are rather slim. The *informational* effect of television is what is thus most important to them. Adults do watch a great deal of television, in many cases more than children do. A 1978 *Psychology Today* cover story described many adults' experience of television leading them into a "frightening world" that causes them to overestimate the number of police in America and their own personal chances of coming to harm as a result of criminal activity.

TV's picture of the world *is* frightening. In spite of the fact that the average police officer goes through an entire career without firing a weapon at anyone, police on television do so three times an hour. Television news, the source of "credible" information on the tube, stresses violent incidents almost exclusively, reinforcing with a "local angle" the lessons of the programs. This message gets through to adults. Recent studies have found other effects, some bordering on the behavioral. A study at UCLA found that adult men who watch prosocial programs are much less likely to behave hurtfully or aggressively around home than those who watch antisocial, violent programs.[11]

It has become an axiom among communication researchers: "What children do to television may be as important as what it does to them..." The tendency is to forget that the viewer is always a participant in the viewing process. Viewers decide what to watch and decide what to do with what they watch. Research indicates that their will in this regard is strongly influenced by the content of the programs they view. Other factors can enhance, influence, counteract, or otherwise affect the messages they receive in content. Certainly, not every child who watches violent programs becomes violent because of doing so. For a certain number of children and for a certain part of each child's behavior, television exposure proves influential. As such, it cannot be ignored either for its impact on the viewer's preparation as a moral being, or for the value society places on its effects.

It is not sufficient, though, to limit investigation of the social and ethical implications of television to consideration of content and its direct effect on behavior. Broader issues are involved in television and telecommunications than their individual effects on viewers. A strong public interest and media reform movement has grown up around advocacy not only for changes in programming service, but also for changes in the regulation and performance of media institutions generally. This movement has as one of its priorities the improvement of service to the variety of needs and interests of the audience, not just the avoidance of potentially negative programming.

As already noted, a holistic perspective is best, one that looks both at programming and at service and, as the telecommunications age becomes reality, also looks at the issues that may arise uniquely there.

The best way to analyze such things is not by having someone else list the implications, prescribe the actions necessary, and predict the outcome. Instead, it is important that individuals become experts in their own right, knowledgeable about these new institutions that now mediate so much of life. Much of the knowledge needed to become so informed comes from evaluation of what has happened in the past. In order to fully understand and be able to ascertain the movements and trends of the future, the present and past must first be understood.

Language becomes a problem. What words are to be used to differentiate between the age that was dominated by radio and television and the new age that will be dominated by new media? Les Brown, the distinguished television editor for *The New York Times* has chosen to call the two phases "TV I" and "TV II." Others call the new age the "age of telematics," or "compunications." In this book, the present age will be called the "television age," because that is the medium that has typified the era for most homes. The future will be called the new "age of telecommunications," or the "information age" because the future will be typified by the introduction into the home of extremely sophisticated systems of telecommunications and information services.

These are exciting times. Before rushing headlong into the future, however, the implications of the current age bear examination.

SECTION II

Television: The Forerunner of the Giant

Introduction

Many television guide magazines carry a feature called "TV Mailbag," a sort of gossip and advice column of the television world where viewers register opinions and ask burning questions. Some of the comments would curl the hair of anyone concerned about such things as televised sex and violence, as the viewers often applaud those aspects of the programs. Running through most of the letters, though, is a consistent thread of complete misunderstanding about how television is paid for, made, packaged, delivered, and changed.

Writers often address a complaint to the magazine, accusing it of somehow being responsible for the cancellation of a program. Writers flail at some unknown entity, claiming that "they" really don't "understand" the viewers. Some letters even criticize characters such as soap opera villians, expecting that the magazine could in some way change the plots of the program.

The ignorance of the television audience about some of the basic facts about the medium is one of the greatest problems facing any who hope for more audience-oriented programming. This problem is at least partially due to television's failure to provide basic education about itself to its viewers. Only recently, for instance, has television yielded to a long-standing suggestion and begun airing the opinions of television critics—a highly controversial practice, at least as far as local stations are concerned.[1]

In discussing the problem of public knowledge of television, mass media observer Dr. William Fore calls some of the most common misunderstandings about television "myths."[2] He describes a number of misapprehensions about television and then sets the record straight. This section will expand Fore's ideas in describing television in terms of common assumptions about it, some of which are true, some false, and most somewhere in between.

An additional aspect of the status of telecommunications in the age of television has been the way Northern Hemisphere media have moved across national borders to pervade the rest of the world, a development that this section will also examine.

Chapter 3

Who's in Charge?

Myth About TV Number One: It is free

This idea about American television is most often cited in comparison with other television systems, such as that of Great Britain, where yearly licensing on sets supports the national networks and BBC television. This argument is also used to compare commercial television here with noncommercial (public) and subscription television here.

Exactly how much television costs is a matter of some controversy, but it should be clear that someone pays for it. America is a capitalist society where, it is often heard, "There is no free lunch."

Society shares the cost of television in a variety of ways. Most people understand that advertisers pay broadcasters for television time and that the money they use is money derived from the sale of their products. But a great deal of controversy surrounds the question of how much of the purchase price of products goes to television, a percentage that varies from product to product. For products that have no clear advantage aside from that which advertising must stimulate, the proportion of their price put into advertising must be higher than for others. For instance, CBS News has reported that Geritol, a heavily promoted iron supplement, is the product with the highest per-unit advertising budget, with 60 percent of its price going to promotion.

Dr. William Fore estimated that for a family of four, about $250 of the money spent on television-advertised products in 1970 paid for television time to promote those products.[1] Since advertising rates have continued to increase, this figure might well have doubled by 1980. Compared with the British annual license fee of 18 pounds (about $35), $500 seems high.

If there were no television advertising, it is not clear that the average family would save money on the products it purchases. One school of thought suggests that the promotion and marketing of competitive products on television brings unit costs down, thus saving consumers money. Although this may be true, many television-advertised products are not truly competitive. The two leading brands of toothpaste, Crest and Gleem, are made by the same company, Procter & Gamble. Procter & Gamble is the largest advertiser in the world, spending more than a quarter billion dollars on advertising in 1979. Procter & Gamble's products dominate television and, with three other advertisers, make up the bulk of products offered there.[2] It could be said that the audience pays for television indirectly, with advertisers actually using the audience's money.

In another way, the audience pays more directly for television: through the purchase and maintenance of television sets and the electricity to run them.

A homespun but insightful Midwest philosopher, a retiree whose days were divided between his television set and his garage workshop, described himself as an early confirmed television addict. He recalled an analogy from his past: When he was young, the medicine show was a frequent attraction in his small town. He described the show as "a couple of mules pulling an old wagon. The show was a guy who must have been a refugee from vaudeville. He sang a little, danced a little, and then, when a large enough crowd had gathered, he'd sell some elixir." Years later when television was introduced, he said he was reminded of the medicine show. He was one of the first people in his town to have a television set, even before there were stations with programs. When the programs finally came, he observed that there were songs and dances by ex-vaudeville stars, interspersed with elixir ads. "Now haven't we come a long way," he mused. "This new technology is just a better way to deliver a medicine show. The difference," he added perceptively, "is that with television, *you* have to buy the wagon."

His analogy is a good one. Just as in the old medicine show, the entertainment on television is paid for by the elixir, not by admission fees of the audience, as is the case with films or theater. The performance of the medium is thus more nearly guided by the needs of the elixir promoter than by the audience. That the audience might wish better entertainment is a tangential issue. As long as the audience puts up with the medicine show and there is no competition, except from similar medicine shows (instead of from something really different), then the situation remains as it is.

Just how much money is involved can boggle the mind. Figures on actual income are hard to come by, but for 1978 the Federal Communications Commission reported that commercial broadcasting had total revenues of $9.5 billion and reported profits of $1.9 billion. Television accounted for 80 percent of those profits. The average 30-second commercial on network prime time in 1980 cost $75,000 to run just once, not counting costs of producing the commercial. The range was from $45,000 for low-rated shows to $150,000 for high-rated ones. For popular specials, the prices are even higher. A 30-second spot on the broadcast of the Super Bowl that year, viewed by 100 million people, sold for $234,000, again not counting production costs.[3]

These figures should serve as a reminder that commercials are not shown on network television without good reason. When advertisers spend 80 percent of the cost of a program segment to produce a single ad, they are sure to have it done *right*. And, if it is to be placed on a program where it costs $75,000 each time it is shown, an ad will have been heavily tested to make sure it is effective or it will have been discarded, a tax-deductible business loss.

Myth About TV Number Two: It is in the entertainment business

This myth about television is probably the one most widely repeated. It is often used by broadcast industry representatives to deny a mandate for television to be involved in activities serving social purpose. For example, at a 1975 conference on children's television, in the midst of a stirring discussion on television's educational impact, an NBC vice-president energetically put forward this myth with the remark, "Now let's not forget what business we are in, after all...we are in *show* business."

Indeed, most people in commercial broadcasting identify strongly with the great American entertainment tradition and see themselves as part of that enterprise. In terms of the earlier "medicine show" analogy, they are, in a sense, part of the modernization of the medicine show's wagon.

However, if the business of television is identified in a market sense, it is emphatically *not* in the entertainment business: Television does not receive its income by entertaining customers, in which case it would be guided by laws of supply and demand in its service of its audience. Instead, television is in the audience-delivery business: Television agrees to deliver to advertisers, in return for payment, the attention of its audience. This is an essential fact: The broadcaster's inventory is not *time*, it is *audience*. Knowing this makes possible some helpful reevaluations of the television experience.

Further, the broadcaster does not sell just any audience—it is a certain quality audience. The trade term for this feature is *demographics*. "Demographics" are statistics on certain characteristics of a population, such as age, sex, income, race, and social class.

In television audience research, the demographic characteristics most often sought and recorded are age, sex, and income. The research firms that analyze audiences (primarily the A.C. Nielsen Company and the American Research Bureau or ARB) report these "demographics" to broadcasters, who use them to price time slots in their television schedules.

These data are important because not all demographic groups are equally valuable audiences for the advertising buyer. The most desirable demographic group is women, ages 18 to 45 or 49. Older, younger, or male audiences are less desirable. Although the situation in the American home is changing slowly, it is still women who watch the most television *and* who control the household purse strings for the sorts of consumable goods most advertised on television.

As measured by these demographics (usually called "ratings") any successful program will attract large numbers of women regardless of its other demographics. An example of the effect of demographics on programming is the case of the "Lawrence Welk Show," a fixture on American television since the 1950s. Amid much public outcry, Welk was canceled by ABC in 1972, not because its audience was smaller than the competition's, but because it was older. In order to sell its time inventory for top dollar, the network wanted to make sure that the people watching were the most attractive to its customers. When Lawrence Welk no longer met that standard, his show was canceled.[4]

A traditional understanding of free market economics explains well the relationship of advertisers to broadcasters. The truism that the buyers—presumably the viewers—can be served well by a free enterprise situation in which they can choose from a variety of sellers, thus making their needs known through market demand, does *not* apply to broadcasting. The real buyers in broadcasting are the advertisers, and *their* demands are first in the minds of television entrepreneurs, not those of the viewers, who are the commodity being sold.

This whole argument about broadcasting became an important policy issue in 1979 when the Federal Communications Commission revealed its intention to "deregulate" radio, allowing the "marketplace" to operate. In truth, the marketplace is already operating in American broadcasting. It has resulted in one of the best, most efficient systems ever known for delivering an audience. It has *not* provided, and cannot provide, a way to satisfy the public's demands. New technologies, to be discussed later, could offer fundamental structural differences toward this end, but they too may fail.

The FCC's proposal is not new. The Commission's reasoning is based on arguments made by "broadcasting-is-show-business" proponents for years, and in legislative initiatives by Congess beginning in 1976. That year the House of Representatives Communications Subcommittee (under Chairman Lionel Van Deerlin of California) jumped on the faddish "deregulation" bandwagon, proposing that constraints on broadcasting were outdated.

During subsequent revisions of the legislation, its backers pointed to experiences in the deregulation of the trucking and airlines industries, where increasing competition has resulted in lower costs for consumers. These situations have no relevance to broadcasting where inventories are strictly limited and will continue to be, particularly without regulation, and where the "customers" are advertisers who are quite pleased with things as they are. The audience—the "public" that broadcasting is supposed to serve—is the commodity it sells. This is the essential curious fact that separates broadcasting from the common understanding of how a business operates.

In this way, broadcasting and some other mass media (magazines and newspapers) are not dissimilar. Most people are aware that a publication's subscription income pays only a small portion of its cost of production compared to the income from advertising. Publishers also describe their circulation in terms of their demographics. In fact, the A.C. Nielsen Company used to be known primarily for its magazine circulation research, not for its broadcast ratings.

An essential difference exists, however, between publishing and broadcasting in this regard. In theory, an unlimited number of publications can be produced, and any reader can subscribe to an unlimited number of publications. The same is *not* true for broadcasting. In broadcasting, the number of stations is strictly limited by the physics of the electromagnetic spectrum. A limited number of stations can operate or they will interfere with one another on the air. Too, the viewer's opportunities to consume quantities of broadcasting are strictly limited by time. The viewer simply cannot consume many broadcasts at one time. (It is curious, though, that in many households two or more television sets and two or more radios are on simultaneously.) This difference between broadcasting and publishing is a creation of the audience research process: While the actual size of a household remains the same, it can be larger for the purpose of counting the circulation of magazines and newspapers, but limited for counting the number of broadcasting experiences it consumes.

Although this difference works out in a number of ways, it is a difference that is gradually disappearing as the technologies of communication converge. Publishing has been heavily influenced by television in its content, but the effects of its gradual assumption of broadcasting's methods of marketing and financing may be still more profound.

Chapter 4

Who's Watching?

Myth About TV Number Three: It isn't that powerful or watched that much

This myth about television is widespread, perhaps because it is hard to believe, based on common experience with the world, that anyone has been markedly changed by television. Data such as the fact that the crime rate has increased along with television use generally have been put forward by alarmists and are easily dismissed by remembering that careful arrangement of statistics can suggest correlations where none exist.

Individual experience also accounts for suspicion about television's impact. Many of today's adults cannot remember a time without television. They look at themselves and say, "I watched 'The Untouchables' and 'The Three Stooges' [two of the most violent programs of all time] and *I'm* not violent . . . What's all the fuss?" Or, "Kids have *always* hit each other . . . Why blame television?"

It is hard to directly answer someone's personal experience with generalized research. For example, a young woman, now in her early 30s, grew up with television. Her parents limited her exposure to violent, live action shows, but put almost no restrictions on violent comedies, violent cartoons, and "fantasy" westerns. As she now reflects, she is not sure how her behavior was affected by these experiences, but she is sure that she knows more about the techniques of violence than her parents would have liked, and finds herself far more tolerant of violence than she wants to be when she considers it.

This woman knows she has learned other things from television. Although her parents did not smoke or drink and wanted their children to be abstainers also, by the time she was in the sixth grade she knew how to buy, open, and light cigarettes and cigars, and the names of various liquors and cocktails, how to mix them, and in what social situations they were considered appropriate. She knew all about the army and war, what boot camp is like, the various ranks in the armed services, and scores of other things that she probably would not have known without television.

A discrepancy exists between what viewers think they know about television's effects on them and what researchers have found its effects to be. This discrepancy has been addressed by a leading researcher, George Gerbner. He comments, "What a viewer learns and what a community absorbs over time are far different things." Gerbner's field of research, which he calls "cultivation analysis," suggests that the television is not important for what it makes viewers "do" or "think" but for what it " . . . contribute(s) to the meaning of all that is done—a more fundamental and ultimately more decisive process. The consequence of mass communication should be sought in the relationships between mass produced and technologically mediated message systems and the broad, common terms of image cultivation in a culture."[1]

Thus there seem to be two different ways in which the power of television can be seen. It can be largely an individual process (How does it affect *me* and my children?) or a larger social process (What are its effects in general?). The latter area is subdivided with the social learning researchers like Bandura and Leibert holding that the larger effect can be seen in terms of individual behaviors, and the cultivation researchers like Gerbner and Gross looking at television's large, subtle effects on all of society.

In all these frames of analysis, television is believed to be powerful. An earlier chapter showed some of the social learning and cognitive developmental findings. Although not all viewers are uniformly affected by exposure to television, some are, and the direction of the effect is consistent with, not contrary to, the dominant messages on television.

Gerbner and others have found similar results on a larger scale. Not only has their recent research suggested that violence remains a fairly constant feature of television from year to year, but also does have effects. Televised violence increases the likelihood that viewers will overestimate the amount of crime in society, the number of police, and, most important, their personal chances of being a victim of crime.[2]

Gerbner believes that the amount of time television is watched is an important determinant of its effects:

We have found that the amount of exposure to television is

an important indicator of the strength of its contributions to ways of thinking and acting. For heavy viewers, television virtually monopolizes and subsumes other sources of information, ideas and consciousness.[3]

The impact of this monopolization tends to erase other social factors that would otherwise make people more divergent in their views of the world. Gerbner and his associates describe this process as "mainstreaming":

> For example, it is well documented that more educated, higher income groups have the most diversified patterns of cultural opportunities and activities; therefore, they tend to be lighter viewers. We found that, when they are light viewers, they also tend to be the least imbued with the television view of the world. But the heavy viewers in the higher education/high income groups respond differently. Their responses to our questions are more like those of other heavy viewers, most of whom have less education and income. It is the college educated, higher income light viewers who diverge from the "mainstream" cultivated by television; heavy viewers of all groups tend to share a relatively homogeneous outlook.[4]

The amount of time people spend watching television is astounding. The distribution of television sets has increased since their introduction in the 1940s to the point that over 98 percent of all homes (79 million) had at least one working set in 1980; 49 percent of those homes had at least one color set. (Radio distribution is even larger in numbers though not in use.) According to recent figures, adults watch two and one-half hours per day, children 12 and under slightly more, and the average set is on nearly seven hours a day. These figures are for average hours per day, seven days a week, 365 days per year. People who respond that they watch less than that must remember that someone is watching *more*, making up for them.

Comparing the amount of time spent viewing television and the amount of time spent on other activities is also revealing. In his 1974 book, *The Responsive Chord*, advertising specialist Tony Schwartz compiled a comparison. He found that each week, people spent more time only on sleeping and working than on television viewing. Listening to radio was a close fourth (26.4 hours for TV, 21.2 for radio). Other "in-home" activities, eating, reading newspapers, magazines, or listening to records lagged far behind (8.4 hours for eating, less than 10 total for the others). "Out-of-home" activities such as films, sporting events, or cultural events (including attending church) totaled less than one hour compared to television's 26! Reading books scored .06 hours (less than 10 minutes per week).[5] Again, these are average figures.

Television is by far the most common appliance in American homes (98 percent is more homes than have running water) and many people think it is a necessity. Court rulings in some states have allowed that a television set is one of the necessities of life that can be kept when a person declares bankruptcy. People who have tried to avoid television often find they cannot. Neighbors, friends, or relatives give them their old sets out of pity. Their children attend school and other social situations so permeated by the language, idiom, and ideology of television that it affects them anyway.

Television *is* powerful in many ways and it is heavily watched. It is important to avoid drawing outrageous conclusions about its effects on everything from soup to nuts, but it is irresponsible to ignore what is known from legitimate experience and data about how it acts and how it is used.

Myth About TV Number Four: The ratings are not accurate

This myth seems to be most often repeated by people whose favorite shows (or the "best" shows) always seem to be canceled. Their belief is based on the feeling that if the show *they* cared about was taken off, it could not have been because their tastes were not universal; rather something must have been wrong with the system by which people were counted. Sometimes the total numbers for shows are good, but they are canceled anyway because the audience is wrong, as was the case with Lawrence Welk, or their time slot pits them against unbeatable shows.

Fraud has happened in the ratings, instances where the numbers have been manipulated, but these all have been related to local situations. Too, certain sectors of society (people without phones or listed telephone numbers, for instance) are systematically ignored by the sampling, but this problem has been corrected.

When viewers learn that the national samples used by Nielsen or Arbitron (the two major ratings firms) are only 1,200 and 1,400 households, they immediately suspect that the problem with the ratings is the small sample. The size of the sample, however, is more than adequate to achieve a level of confidence that the figures are 95 percent correct. To achieve even 96 percent correctness, the size of the sample would need to be doubled, at the least, and the firms and their clients, the broadcasters and advertisers, believe that the expense involved would not be justified.

Sampling techniques vary from study to study, but ratings are gathered in two basic ways: the logbook, and "the box." The logbook is mailed to sample households along with instructions asking them to record their viewing in quarter-hour periods for a week, including who watched and how long. "The box" is attached directly to the set. It either reports directly how long the set is on and to which channel it is tuned or records that data in coded form and is checked periodically by a technician. Boxes are attached for longer periods of time than logs are in use, but the sample is continually rotated with new households joining and old ones being removed.

Additional samples are taken by telephone to cross-check the other two systems, which cross-check each other (the box can verify which channel is on, and for how long, facts the viewers could misrepresent, but it cannot record who was watching, the important demographic information).

An additional sampling tool is the "sweep week." "Sweep week" occurs once each season, usually in November, February, and May, and more detailed ratings, both in the kind of information gathered and in the numbers of cities where information is gathered, are taken. During these "sweep" periods, the networks and local stations put on their best "stunts" to attract the greatest audience. These sweeps, more than anything else, determine the pricing of network and local programs for the next season. One industry representative estimates that one rating point gained or lost in the sweeps can mean $7 million in revenue to a network.

Rather than being built on hard, unchangeable data (the impression given by all the facts and figures), the ratings game is built on a complex set of agreements and understandings among advertisers, broadcasters, ad agencies, and rating firms. The "sweep week" is an example. The viewers' interests might be served by surprise "sweeps" that would keep the industry constantly vigilant to the quality of its offerings, never knowing quite when it might be "checked up on." Instead, *everyone* knows which weeks the ratings are being taken and broadcasters save their best offerings to throw at each other then. The rest of the season can seem dull by comparison.

Another interesting example of this phenomenon was the HUT controversy of 1978. HUT (Homes Using Television) figures for that year showed for the first time a *decline* in overall television use—a very threatening situation if true. Nielsen reported that year that the decline was probably related to an adjustment made in the sample to include more childless homes—a growing trend in society—and those homes tended to use less television than others. This made perfect sense from the perspective of a researcher but it was a disaster for broadcasters who had to try to make the same or more income from ads watched by fewer people. A hastily called conference at Nielsen's Florida office settled the matter with an agreement but the experience revealed how much the ratings game relies on consensus rather than on hard data.

Ratings points can spell disaster for shows, careers, investors, and viewers. Ratings are, in fact, reported in two ways. One, the *rating,* reports in millions the numbers of viewers or households tuning in a certain show. The other, the *share,* reports the percentage of the viewing audience at that time that was watching a particular program. Both figures are needed to get a good idea of a program's audience. For instance, a popular program of the 1970s, such as "Laverne and Shirley," might have garnered a *rating* of 33.5 (33.5 million viewers) compared to 20 for Johnny Carson's "Tonight" show. Carson's was the more successful program, however, because the share for his show might well have been 80 to 85 (the percentage of people watching at the late hour he was on) compared to 35 to 40 for "Laverne and Shirley."

Shows are often canceled, bought, or sold based on *one* or *two* rating—and share—points. (Remember the $7 million rating point mentioned earlier.) The 1980 spring sweep found CBS to be ahead of rival ABC overall by a *tenth* of a rating point (19.5 to 19.4), a fact that added significantly to the popularity of its stock and its advertising time.[6]

These few points, or fractions of them, are vital to broadcasting, yet there is only consensus agreement on their validity. This is because differences this small fall well within the margin for error of the sampling technique used.

So the ratings are fairly accurate. They are accurate enough that their findings about wide variations in audience for most shows can be trusted. They are correct in finding that religious programs are not very popular (even the successful ones!).

Broadcasters often say to critics, "Look, we know what people want, we have the ratings." The accurate response is that this is *not* what can be known from ratings. The ratings report, fairly accurately, what people *watched* last night, not what they wanted to watch, not what they needed to watch (a concern of the churches and the schools), not even what they would have watched if they had had a real choice, only what they watched last night, last week, or last month.

There is an overall lack of differentiation among the networks. Anyone unaware from other sources which network was on would have a difficult time correctly identifying the network merely by tun-`ing in a show. One advertising executive called the 1980 season "imitative," a kind word for the same idea.[7] Using the ratings system to justify the content of American television is as if the American people were taken to lunch and allowed to choose among Burger King, Burger Chef, and McDonalds, after which it was announced that they all liked *hamburgers*.

The ratings *are* fairly accurate. What they reveal is only a limited amount of the information that might be helpful in developing a broadcast system based on priorities other than the maximization of profit for entrepreneurs.

Chapter 5

Through the Screen Darkly

Myth About TV Number Five: There are altruistic sponsors

If altruism means sponsoring programs purely for public benefit, expecting nothing in return, this is another myth. *Nothing* is advertised on television without reason. A corporation might support quality programming because it is interested in the welfare of society, but the corporation would also let the viewers know of its sponsorship. (There is nothing necessarily wrong with such activity, but viewers should be aware of it.)

Corporate advertising that is "image" rather than "product" oriented is often called "institutional advertising." It is handled carefully by networks in some cases so as to avoid manipulation by advertisers. One such case has been CBS' long-standing refusal to allow Mobil Oil Company to place institutional ads to rebut CBS News' coverage of oil company profits. In 1980, out of frustration, Mobil agreed to sponsor a new national news program, "Independent Network News," where its ads could appear.

A curious case of "altruism" is reported by Eric Barnouw in his book, *The Sponsor: Notes on a Modern Potentate.* He noted that International Telephone and Telegraph Corp., sullied in the early 1970s by its involvement in the CIA subversion of Chile's Allende regime, sought to buy back some of its good image by sponsoring a children's program called "The Big Blue Marble." ITT then ran prime time television commercials to promote the program. Barnouw concludes:

> "The Big Blue Marble" films cost $4 million to produce but were *given away* to television stations, with the result that they appeared mainly in fringe periods. Not so the corporate commercials about "The Big Blue Marble" series and the other intercultural

good deeds. These commercials appeared in prime time at a cost of $4.2 million in 1974, and of $3.7 million in 1975.[1]

Similar promotion activities have surrounded public television specials. Corporate underwriting of public broadcasting, in itself a form of sponsorship, got a boost in 1980 when Congress agreed to allow a certain amount of institutional advertising to appear there. What had seemed to be altruistic support of certain programs by large donors in exchange for a brief mention was, in fact, a legal constraint that has now been partially lifted.[2]

Myth About TV Number Six: It is a "mirror" of society

This myth about television often crops up in arguments by apologists both inside and outside the industry. The argument that criticism of news performance is "killing the messenger because the message is distasteful" is one form this argument takes. Another is the comment heard many times: "You know why there is violence on television, it's because there's violence in real life." One college student put it this way: "So what's the problem with kids learning to be violent from television? It's just preparing them to go out and face life." When asked how he thought physical violence related to life, he responded, "So when kids grow up and get drafted and go to Vietnam they will know how to save their own lives." This conversation took place long after the war there had ended so the questioner assumed the student was a Vietnam veteran. "No," he said, "but I saw it on TV."

So is television a mirror? There is no way that television can act as a mirror of the world. The reason is basic to the television production and viewing process: There simply is not time for all of reality to be presented (who would seriously *want* it to be?); selections have to be made. The selection process involves hundreds of different and interrelated choices made by the producers of television between the gathering of sound images and their ultimate consumption by the audience.

Consider the situation in television news, which is in many ways unique but is considered by most people to be the most "credible" or "real" aspect of television presentation.

A news crew assigned to cover a given story goes to the location where it is unfolding. Were they actually to serve as only "eyes and ears" for the audience, they would set up the camera and microphones, turn them on, and leave the images on until the event ended. A 1980 experiment with no play-by-play coverage of a professional football game used just this approach—to mixed reaction. On a scale with "total reflection" on one side and "total control and manipulation of images" on the other, this scenario would be near the "reflection" end.

"Mirror Metaphor" Continuum Condition 1

Total, passive reflection is not possible. The very presence of news crews and cameras changes the course of events covered. Other factors inherent in the medium of television also intervene to make this "near reflection" less total. The first is time. Most live or taped news stories on local stations are allowed less than 60 seconds, so only the best scenes (called "cover" shots) are taken with "representative" snippets of sound. The short scene must be introduced by an anchor at the station or the reporter covering it. What the reporters say about the scene, what the camera people shoot, what the film or tape editor leaves in or leaves out, how the anchor introduces the story, where it is placed in the show, whether it is placed at all, indeed, whether it is even *covered*, move the point on the scale farther and farther from pure "reflection."

"Mirror Metaphor" Continuum Condition 2

News distortion is not a matter of controversy. It exists. The only questions to be asked are "Who distorts?" and "How?" The same phenomena are true of newspapers and news magazines. The differences are that more people get more of their news from television than from any other source and news on television is far more restricted in content because of time than in most other sources. The entire "CBS Evening News," including the ads, would fit on the front page of *The New York Times.*

The selection process in news programming is similar to the selection process that occurs in the rest of television. A primary difference is that there is more freedom in the rest of television than in news about the "scene" to be reflected. Thus the whole of entertainment television is farther along the scale than is the whole of news.

But does television not "reflect" society in a certain sense? Hermenio Traviesas, the distinguished former head of NBC's censorship department, was never so naive as to suggest that television accurately reflected the world in detail, but he did see his role as seeing that television was not too far out of step with America: "When we see the mood in America changing . . . for instance, about sex . . . then we change, too."[3] Another network official put it slightly differently: "Our job is to stay with the tide of American taste, but we have a responsibility to lead *just a little bit."*

Certainly these sentiments and others of the broadcasting community carry the sense that the enterprise is not so much a matter of "mirroring" as it is one of staying consonant with the broadcasters' perception of American taste, often derived from the mail they receive.[4] This consonance with taste can yield to a number of pressures, most often the desire for better ratings. Television derives most of its direction from the ratings game, and this forms another pressure on the "selection of content" scale. This pressure can even contravene artistic impulse, such as in the case of the 1977 series, "James at Fifteen." This much-acclaimed program concerned a boy's adolescence and was uniquely and sensitively done. When the writer of the series had James turn 16, the network decided that he should enter adulthood by having sexual intercourse on his birthday with a schoolmate. The writer quit the series over this infringement on his artistic prerogative, and James' adulthood was consummated on network schedule.[5]

Another better known example of this phenomenon was the 1977 resignation of Esther Rolle from the "Good Times" series in which she had starred as Florida, the mother and protagonist. The series was one of the few on the air with primarily black characters, a black family living near the poverty line in a public housing project. When the character who played the father left, he was not replaced, and the family was refit into a more stereotypic "welfare" situation.

Rolle's frustration was over the characterization of the older son, J.J. (played by Jimmy Walker). She explained:

> I just couldn't go on any longer with that kind of outrageous character as the role model for black children . . . he didn't work, he cut school, he was disrespectful . . . [6]

The producers and the network knew they had a winner with the J.J. character, so he stayed and Rolle left.

An issue related to the question of distortion by television is the question of censorship of television. Censorship is most often mentioned with respect to government pressure on television news. Spiro Agnew's attacks on network news and network response to them have been well documented. But the fascinating and forgotten fact is that most censorship is done by the *networks* who are in no way accountable to the public for their use of this power. While outright censorship in the high school civics sense of the word is frightening, it seems odd that a society that at least considers itself to have representative government leaves this important task to a few private institutions. These institutions engage in censorship in response to pressures from many sources, including the public, but primarily in response to advertiser needs.

So television does not mirror society. Other metaphors have been used to describe its function more accurately. John Murray, who worked on the surgeon general's study in the 1970s, put it this way.

> Television can be considered to be the window on the world, a school, if you will, through which the child first perceives society and then learns from repeated example to cope with the vicissitudes of living. [7]

George Conklin, a widely respected media educator from the Pacific School of Religion, says television is "a carnival mirror, which reflects back to us images that seem to be right and true, but which are skewed a bit." [8]

Instead of reflecting society, television selectively represents society to itself. Producers, directors, advertisers, the networks—all are involved in choosing which scenes to show and which to avoid, and how the chosen images are to be presented. Some selection process, of course, is necessary. No one would want television to be a mirror.

People who criticize television for its lack of balance, in the representation of women and minorities, for instance, are often accused of wanting television to be a mirror, to offer a bland, purely representational view of society. But this is not their goal. The primary goal of pointing out that the images on television are out of kilter with reality is not to bring about a mirror image of society. This criticism is intended to focus the attention of the television institutions on issues other than ratings, placements, and letters.

No responsible television critics wish to set themselves up as censors, a charge often leveled at them. Knowledgeable people, on the other hand, cannot help but notice that censorship is already a fact in broadcasting. It is only reasonable to expect that some censorship be responsive to positive social goals, not only reactive to the threat of lower ratings or public outcry.

Chapter 6

Airwaves in the Fog

Myth About TV Number Seven: The FCC regulates broadcasting in the "public interest"

Anyone who writes to the Federal Communications Commission complaining about a television program is likely to receive a polite, firm, and surprising reply: "I'm sorry, but we don't have anything to do with content of programs."

The Communications Act of 1934, which governs American broadcasting, contains in Section 326 the following words:

> Nothing in this act shall be understood or construed to give the Commission the power of censorship over the radio communications signals transmitted by any radio station, and no regulation or condition shall be promulgated or fixed by the Commission which interfere with the right of free speech by means of radio communication.[1]

While review of content is not necessarily censorship of it, the Commission has tended to interpret the Section 326 prohibition very broadly and has avoided totally consideration of content.[2]

Other areas commonly assumed to be under specific FCC scrutiny often are not. The following list is a guide to which areas of broadcasting do and do not come under FCC scrutiny.

Commercial practices. The Commission has no regulations directly covering commercials except prohibitions against "loudness," in cooperation with the Federal Trade Commission, certain aspects of deception, and prohibitions against blatantly subliminal advertising.[3]

The FCC does *not* have rules governing the length, number, or type of commercials carried. The FCC has tended to be concerned only with the number of ads but has no formal rules on the subject. What regulations do exist are industry self-regulations administered by the code of the National Association of Broadcasters (NAB). The code's limits on the amount of advertising are the ones generally followed by the industry and accepted by the FCC. The Commission has proven to be mostly interested in following whatever limits the code established.[4]

The Commission has *not* exercised any control over the length or content of commercials. Length has been entirely a matter of industry discretion and varies greatly. Whereas the 60-second ad used to be the rule, the 30-second ad has now become the standard (stuffing more ads into the same amount of time) and the "piggyback" (two 15s in a 30-second slot) and "triggyback" (three 10s in a slot) have been tried.

The content of commercials and the kinds of products that can be advertised are also matters of NAB code discretion only, not the FCC's. For instance, hard liquor cannot be advertised on television (only wine and beer—a curious situation since the programs themselves model liquor consumption more effectively than ads ever could). Originally, feminine hygiene products could not be advertised, and when they were introduced on television, they were limited to late evening hours. Contraceptives cannot be advertised on television, although there is some pressure for this to be allowed, a move that is opposed by many groups.[5]

Specific kinds of educational, religious, or public affairs programs. The FCC has no formal requirements for these other than a statement of general guidelines called the *1960 Programming Statement.* Before 1960, the FCC had requirements of a sort, including one that a certain amount of the public affairs programming be broadcast "free." That requirement no longer exists. The FCC does not care whether public affairs programs, which often deal with controversial issues and thus are not popular with advertisers, are sponsored or not. The effect is that stations are likely to choose "softer" public affairs programs—ones more palatable to advertisers in order to sell them.

A parallel situation has occurred with religious programming. Religious programs that can pay their way (usually the ones by independents who solicit contributions on the air) get more favorable treatment than programs that are put on as a "public service."

"Religious" and "educational" stations. The Commission recognizes only two classes of stations, "commercial" and "noncom-

mercial." Most educational stations, primarily training stations at educational institutions, are "noncommercial" stations by Commission definition. Most "religious" stations are commercial.

The regulations governing commercial and noncommercial stations are surprisingly similar. The primary differences are in the initial license fees ($50 for noncommercial versus $2,000 for commercial) and the fact that noncommercial stations cannot sell ads. Noncommercial status was intended by the Commission and the Public Broadcasting Act of 1967 to be an opportunity for public alternative services to be started.

The phenomenon of religious broadcasting is a relatively new and interesting one. Many so-called "religious" stations claim for themselves a special distinction and a special set of mandates when, in fact, they must meet the same requirements as any other commercial station. Some evidence indicates they are *not* meeting these requirements and that the Commission is allowing a special privilege in this class of station.

This issue came to a head in a 1975 petition for rule making (RM—for rule making—2493, otherwise known as the Lansman-Milam petition). This petition, a legal document filed with the FCC, asked that the Commission evaluate the practice of issuing noncommercial licenses to religious or other sectarian groups. The petitioners reasoned that although the Commission intended for these noncommercial licenses to be available at lower cost for noncommercial purposes, the religious and other sectarian groups using them were using them to make money.

The Commission received an outpouring of mail about this issue even though it sent back the petition later the same summer. Even into 1980 the Commission continued to receive hundreds of letters a week from concerned people motivated to write by religious broadcasters who had told them that this was an attack on religious broadcasting.

The FCC interest in exerting more control over TV

In his analysis, William Fore suggests that the FCC will never move to a more comprehensive regulation of broadcasting because it is too closely tied to the interests of broadcasters.

> Every Congressman has two very important constituents. His local newspaper editor and his local radio station manager. Every time the Commission puts the heat on broadcasting, Con-

gressmen start getting calls from broadcasters back home and the next thing you know, Congress decides to hold hearings on the FCC's budget, and the commissioners are dressed down in open session. Every time that happens, commissioners and staff get more reluctant to antagonize the industry.[6]

Barry Cole and Mal Oettinger, longtime Commission watchers, point out in their 1978 book, *Reluctant Regulators,* that the Commission responded to direct lobbying by broadcasters as well. They quote a Brookings Institution study:

> It is obvious that the president of a broadcasting station in a town of 25,000 will have no trouble seeing a commissioner; whereas in some cases even a division chief or a bureau chief (Commission senior staff people) may have to wait for several days to see a commissioner. The current priorities of most of the commissioners seem to be: (1) parties with direct economic interests; (2) Congress and the administration; (3) the Commission staff; (4) outside persons with public interest but no economic interest (outside public interest groups and individuals).[7]

Cole and Oettinger point out that the situation has changed markedly since 1968 with two or three developments accounting for a broader range of interests being heard at the Commission. They point to the WLBT decision in 1978, which will be discussed later, and the formation of two new industry lobbying groups as important in this change.

In 1975, radio broadcasters formed the National Radio Broadcasters Association. They acted from their dissatisfaction with what they believed to be undue emphasis on television's interests by the National Association of Broadcasters. By 1978, the NRBA had become a lobbying force significant enough to cause serious dissension in the broadcasting industry over communications legislation. The other industry development was the strengthening of the National Cable Television Association (NCTA), which paralleled the development of cable as an important service. Cole and Oettinger also cite increasing pressure on the Commission from other agencies of government, especially the Justice Department as it attempts to reduce concentration in media ownership.

> Formerly a broadcaster who opposed something and could not get satisfaction from the FCC could go to Congress and exert

the necessary pressure. Now, however, Congress, as well as the FCC, is getting pressure from other sources. These other sources include citizens groups and nonbroadcasting industries which, with their political clout and financial resources, are sometimes more influential.[8]

Most of the available evidence suggests that the Commission lacks broad authority in areas of television and radio content. At least its authority is much narrower than has often been thought by viewers and the public.

The climate for communications regulation underwent a substantial change in 1966 when a Federal Court handed down a decision in the WLBT case. Until then, only those companies or individuals who had had a direct economic interest had been allowed to participate in FCC proceedings in the licensing, transfer (sale), or license renewal of stations.

The Communications Act of 1934 had established that the airways were "a public resource" which, because they were limited (only so many stations can be on the air), should be licensed to the best applicants. In the language of the Act, these licensees should act as "public trustees" and should provide "outstanding service" to the "public interest, convenience, and necessity." In order that this take place, the act established the FCC as an independent federal agency to oversee technical and other performance of broadcasters, radio operators, and telephone companies.

The Act intended that the Commission represent the public in these proceedings; thus no provisions were made for public input. The tendency, however, has been for the Commission to take counsel with industry representatives and others affected economically by FCC decisions. As was pointed out in the WLBT decision, the FCC assumed that as long as people watched and listened, they were being served.

This situation changed as a result of the increased scrutiny of all social institutions in the 1960s. In 1964, four parties with no economic interest asked the FCC for permission to participate in the license renewal of station WLBT-TV in Jackson, Mississippi. Their complaints concentrated on the station's lack of service to the black community.

The parties who petitioned the FCC included the Office of Communication of the United Church of Christ, which became the primary litigant when the FCC refused to grant them "standing" (the right to participate). The church appealed to the Federal District

Court of Appeals, and in the case *United Church of Christ v. FCC*, established that FCC practices had ignored the interests of the public and that public groups could be "parties in interest" to FCC proceedings.

Since 1968, therefore, comments from public individuals and groups are accepted by the Commission and, more important, the public may file "petitions to deny" at license renewal time.

Each television or radio station is licensed for three years; most reapply for renewal at each expiration. Each application ostensibly is the same as the first, and there can be no presumption that relicensing will occur. In practice, few licenses are challenged by other applicants; thus, few renewals are denied.

In effect, licenses have been perpetually renewed, and positive action to deny those licenses has been necessary for even the most irresponsible service to be ended. Petitions to deny have not fared well, however. Of the 500-odd petitions filed against renewals in the 10 years ending in 1979, only two were successful.[9]

As the legal record governing petitions has developed over the years, a narrow range of bases has emerged on which petitions to deny can be filed. In principle, the Act's provision for petitions, and the court's action in the WLBT case, should have encouraged input to the FCC along a broad range of license performance issues. This has not occurred. Instead, a restricted set of criteria has emerged. This is based on a number of key rulings which held that petitions must raise specific allegations that, if true, would show that granting the license would obviously be "inconsistent with the public interest."[10]

This record severely limits the bases available to petitions in the programming area, the aspect of broadcasting of most interest to the general public. Petitions that have the greatest potential for success are based on failure of the station to follow FCC procedures in preparation of applications, failure to program to meet community needs listed in vague "ascertainment" documents, and violations of the FCC's special Equal Employment Opportunity rules. Even where these bases are used, it is obvious from the small number of licenses actually denied that it is unlikely the Commission will act.

This is not to deny the effect of the development of public standing. The potential power to petition has resulted in a different sort of attention by broadcasters to their communities, often bringing about agreements for change made between interested local groups and broadcasters. One public interest attorney who has been active in this area reflected on the issue:

> Some important elements of citizen-licensee agreements
> have been dishonored with time, but what changes *have* been
> maintained, have been because citizens' groups continue to be a
> reminder that there is a means of redress of grievance—the peti-
> tion to deny.[11]

Nor can the effect of the WLBT decision on the Commission be
evaluated only on terms of how many licenses actually have been
denied. One public interest activist put it this way:

> Prior to the WLBT decision in 1966, the FCC heard almost
> exclusively from representatives of the telecommunications in-
> dustries. These industries possessed (and continue to possess) the
> high degree of involvement, economic strength, and organiza-
> tional cohesion to present their views to the FCC consistently and
> coherently. Since 1966, the participation of citizen groups has
> provided the Commission with a greater range of ideas and infor-
> mation than the communications industries alone could have of-
> fered. The addition of alternative viewpoints has made it more
> likely that Commission decisions will serve the interests of the
> general public.[12]

Can the FCC regulate broadcasting in the public interest? The
answer is difficult, but according to most evidence, its record has not
been good. Even its handling of the development of the new technology of cable TV has been typified by broadcaster-oriented regulation, which has denied the Commission a significant role in shaping
this important new medium. It appears that the Commission's role in
new developments will not be as an active proponent of the public's
interest over against those of the telecommunications industries.

Myth About TV Number Eight: When the new technologies are in place broadcasting will go out of existence

Subsequent chapters will deal with the emergence of the new
communications technologies of cable, satellites, fiber optics, home
video, and home computers. The coming revolution has much in
common with earlier revolutions in communication such as the
emergence of modern telecommunications with the advent of
telegraphy, electricity, and broadcasting. One thing the coming and
earlier revolutions have in common is the sweeping claims for their

potential. Among the varied predictions of things this new age will bring are "home information centers" which will bring all the resources of the phone book, local newspaper, library, and banks to the home; decentralization of the workplace, allowing people to work from home; 200 to 300 channels of cable programming of entertainment and education; and a total reorientation of social institutions such as the church and the schools to account for this new reality.

Almost all these scenarios give attention to the impact that these new technologies will have on conventional broadcasting. All the new technologies will use the television set in some way. As new uses place a demand on the amount of time a set can be on and in use, most people assume that conventional broadcasting will suffer as people spend more and more time using their sets for things other than watching commercial television, particularly network television.

Initial research on this issue indicates that for the time being, the networks' fears are somewhat unfounded. In cities where cable television and other technologies bring in channels that cannot be seen otherwise (the greatest risk of "competing for time" with network TV) the homes that have cable do not seem to watch less conventional television. Instead, their total viewing increases, with their consumption of "alternative channels" adding to their regular viewing. For instance, one of the new technologies, home video recorders, allows people to tape programs while they watch another channel or while they are away from home. The viewing of these taped programs must take place in addition to regular viewing. Another pressure on these households is that, having spent $700 for a recorder or $15 to $35 a month for cable service, these things should be used in order to justify the expenditure.

When cable penetration (the percentage of total U.S. households with cable) passes a certain point, probably 30 percent, its effects on network television will become more pronounced. Most observers assume that at some point, the total amount of time spent with television in the average home will reach a limit, and the proportion of it spent on traditional network television will begin to decrease.

The networks themselves are becoming aware of this fact. By the end of 1980, all the networks had major corporate divisions devoted to marketing programming exclusively to cable television. It will be quite some time before the networks as they are now known cease to exist. By then, these corporations will have found an important niche in cable television. Other kinds of "networks" will have formed around speciality programs on cable. But the hallmark of the traditional network will be gone—the contractual arrangement with each outlet or station binding it to carry only network-originated films.

The interest of the commmercial networks in cable is not new, but their willingness to cooperate with and capitalize on cable must be taken as a tacit admission that their early resistance to cable's growth has now given way to a more rational, pragmatic approach.

Even the most dispassionate observer would recognize that for the first 20 years of cable's growth, the three commercial networks and the commercial stations were dead set against it. Their effectiveness at advocating with the FCC was largely responsible for the Commission's consistent reluctance to allow cable to grow to its potential.[13] The Commission's earliest rules saw cable only as "signal enhancement service" and subsequent regulations sought to limit cable's ability to bring signals other than those already available in the community. It has only been since court rulings voided the FCC rules that growth in cable and cable service has occurred.

Will conventional broadcasting as it is known cease because of cable and the other technologies? It depends on what is meant by "conventional broadcasting." If that means the strict system of commercial broadcasting based on local affiliates of traditional networks, then yes, that system will gradually fade from prominence, probably between the mid-1980s and mid-1990s. If conventional broadcasting means the quality and quantity of entertainment programming that has been typical of the three networks, the future is less clear. The most probable scenario is a proliferation of such programming to fill the increased capacity of the cable spectrum. The networks' cable divisions are preparing to provide that kind of programming to cable. As cable gradually changes from a subscription income service to a subscription advertising service, the "least common denominator" will probably once again hold sway. The difference will be that the "least common denominator" audience for cable will be a higher income one than for "free" TV and the services that conventional broadcasting traditionally provided for the lower income groups might no longer be available to them.

Chapter 7

The Bully of the Global Village

> The media are about politics, and commerce and ideas. This
> is a strange enough combination even when the media stay at
> home. But as an item of international trade the combination is
> even more unusual. When a government allows news importation
> it is in effect importing a piece of another country's politics—which
> is true of no other import. The media also set out to entertain and
> intrigue—to make people laugh or cry—they have an emotional
> appeal unlike other products. And because the media also deal in
> ideas, their influence can be unpredictable in form and in
> strength.[1]

A California couple spent several months in 1974 backpacking
throughout southeast Asia. They intended to steer clear of modern
civilization as much as possible, choosing instead to tour rural areas
of Laos, Borneo, and Nepal. They were shocked to discover that
they could have kept up with the entire season's episodes of "Streets
of San Francisco" by watching the community television set in nearly
every little hamlet they came to. That was not all. They had wanted
to have as little impact as possible on the cultures they encountered,
not wanting their Western ways to encroach on the local ways of life.
Still, they found their jeans and Frisbee (the two items of Western
goods they had with them) were sources of constant envy, at least on
the part of the youngsters in the villages. They had seen jeans on
American television shows and wanted them, too. They were also
very interested in the image of America they got from the "Streets of
San Francisco," "Ironside" and other such programs and questioned
the couple about how many cars they owned and whether they carried a gun with them to feel safe on the streets.

This couple's experience is a microcosm of the effect that the exportation of American television has on recipient countries. While the sort of misperceptions represented by the questions they were asked was probably not terribly serious (except as it reveals perceptions of the U.S. by the rest of the world), the effect of such programming on those cultures is a serious matter. Roy Nehall, a church and social reform advocate from the Caribbean, has called the effect of such exportation "cultural genocide."[2] His thinking about the situation in the Caribbean is supported by information from other places. It was reported in 1972 that the Communist Party Central Committee in Poland had changed its weekly meeting from Wednesday to Thursday night because "Dr. Kildare" was on Wednesday nights. The 1980 airing of "Rich Man Poor Man" in Poland had a similar impact. Other recent reports confirm the general trend of American television's dominance.

Another phenomenon of 1980 was the worldwide interest in the evening soap opera "Dallas." That program took Britain by storm (it was carried by the public network BBC) with the spring 1980 shooting of a major character, J. R. Ewing, and the fall 1980 revelation of his assailant achieving some of the highest audiences in BBC's history. (The U.S. ratings for those segments were historic highs as well.)

What impact do such incidents have on the societies where they occur? Surely British culture is not heavily affected by such instances, but what of societies far less Westernized?

The U.S. exports more programming than any other country, far outstripping any rivals. One explanation is the economics of commercial broadcasting: Producers make very little of their income from first-runs of their shows on network television. Instead they rely almost entirely on domestic and international syndication for their return on investment. Once purchased, U.S. programs can be distributed to other countries for much less money than domestic production would cost there. The effect on local production is that although most countries have national television systems, very few of them produce the majority of the programs their people watch.

This situation of U.S. dominance is not new nor is it unique to television. In his 1975 book, *The Media are American,* Jeremy Tunstall explains how the American film and television industry attained dominance over the world:

> Alan Wells elaborates how American television imperialism works in Latin America. Latin America television, since its birth,

has been dominated by United States finance, companies, technology, programming—and, above all, dominated by New York advertising agencies in practice. There is a very substantial U.S. direct ownership interest in Latin American television stations. World Vision, an ominously titled subsidiary of the U.S. national ABC network, plays a dominant role in Latin America; American advertising agencies not only produce most of the very numerous commercial breaks but also sponsor, shape and determine the whole pattern of programming and importing from the U.S.A. Indeed, "approximately 80 percent of the hemisphere's current programs—including the 'Flintstones,' 'I Love Lucy,' 'Bonanza,' and 'Route 66'—were produced in the United States." This near monopoly of North American television programming within South America distorts entire economies away from "producerism" and towards "consumerism." Madison Avenue picture-tube imperialism has triumphed in every Latin American country except Cuba.[3]

In *Mass Communications and American Empire,* Herbert Schiller describes in detail how the American film, television, and popular music establishment has come to operate as an enterprise almost independent of national sovereignty and political direction.

> The American Broadcasting Company, the third major network company in the United States, has been the most active in the international field, perhaps compensating for its somewhat less influential position in the domestic market, where it ranks behind CBS and NBC. ABC has organized an international TV network, World Vision, which, at last estimate, can reach 60 percent of all world television homes where sponsorship is permitted (a total of 23,000,000 TV homes). In the 26 nations where World Vision operates, ABC has some financial involvement in telecasting in the following countries: Canada, Guatemala, El Salvador, Honduras, Costa Rica, Panama, Colombia, Venezuela, Ecuador, Argentina, Lebanon, Japan, Ryukyus, Philippines, Australia, Chili, and Bermuda.[4]

Cees Hamelink has drawn this issue into a broader context of mass communications' place in the establishment of the dominance of transnational corporations which, while they certainly are using media to serve their institutional needs, still affect the cultures and societies of recipient countries.

> There is a *communication industrial complex* which repre-

sents the dialectal linkage of three vital elements of global economic power: control of supporting capital, control of technology
(its development, application and transfer), and the control of
marketing mechanisms. This complex controls the centralized international communications' structure. The function of this communications' structure is the promotion of messages which represent a configuration of sociocultural values pertinent to vested
economic interests. This value configuration tends to conflict with
the sociocultural value configuration which represents public interests.[5]

International debate over the issues involved has recently focused on the flow of information, meaning entertainment and news.
This focus on entertainment recognizes the implicit problem illus-
trated by the experience of the American couple in Asia. Entertainment programming can be a powerful crystalizer of ideas and social-
izing influence, even though its intent may be neutral.

This debate has focused around the New International Information Order (NIIO), and encompasses a call by the Third World and
non-aligned nations (with a variety of support from other interests,
Eastern and Western) for a greater role in determining the shape of
information service inside and outside their borders. The issue was
first raised as early as 1970, but found its most prominent voice in the
mid-1970s in the person of Mustapha Masmoudi, the Tunisian am-
bassador to the United Nations, who pressed the issue repeadly in
that forum. In 1977, the United Nations Economic Scientific and
Cultural Organization (UNESCO) empaneled a commission chaired
by Sean MacBride, a widely respected Irish internationalist. That
report, adopt by 154 member nations of UNESCO in October 1980,
attempted to plot a middle ground between the most extreme sides
of the debate.

First of all the criticisms formulated in many developing countries, reiterated by socialist countries and supported by many researchers and journalists in western countries, start from the observation that certain powerful and technologically advanced states
exploit their advantages to exercise a form of cultural and
ideological domination which jeopardizes the national identity of
other countries. The problems raised by one-way information flow
and by the existence of monopolistic and oligopolistic trends in international flows have been widely discussed in many international
instances, gatherings and seminars. It has been frequently stated, in
particular, that due to the fact that the content of information is

largely produced by the main developed countries the image of the developing countries is frequently falsely distorted. More serious still, according to some vigorous critics, is this false image, harmful to their inner balance, which is presented to the developing countries themselves. The dangers and fears created by the potentialities of direct satellite broadcasting stimulated demand for a balanced flow of information. It was when these questions first came to be discussed that increasing anxiety arose about the content and quality of the information transmitted, together with a dawning awareness of the lag in developing countries in news production and transmission.

On the other hand many media professionals consider that, while the existence of these imbalances and dangers which they entail cannot be denied, stressing the one-way news flow can lead to further restrictions on the freedom of information and to strengthening the hand of those in favor of reducing the inflow of information; the consequence would be a radical break in the concept of free flow. Assuming that the diversity of opinions, news, messages and sources is a precondition for truly democratic communication, this school of thought has also considered that the "decolonization of information" must not serve as a pretext for bringing information under the exclusive control of government authorities, and thereby allowing them to impose their own image of reality on their peoples.[6]

What are the issues here and why are they so controversial? To people in the Northern Hemisphere, the idea of a "free flow" of information makes the most sense. It is a cornerstone of the democracies and has served well. In practice, however, the free flow, allowing people to decide what they will believe, does not work out the same way for less developed countries. Because of the North's immeasurable advantage technologically and economically, its "free" media dominate their media to the disadvantage of the less developed countries. Since those "free" media have the attention of the entire world, their coverage of the Third World is about the only coverage it gets.

Any observer of American network television news must realize that "newsworthy" events in the Third World are few, limited pretty much to catastrophes and revolutions. Even the venerable *New York Times'* interest in minor countries is limited to small items about bus accidents and the like.[7]

One simple indicator of television's interest in the rest of the world comes from Edward J. Epstein's major study of NBC news.

He found that the prominence of a news story in any newscast is clearly related to whether there is film on the story. Film or videotape stories are carried by the American networks only if they originate from their own news crews. Many other sources for news films, such as BBC, cover the Third World much more carefully than do U.S. networks, but their film, which is available to NBC, is never used. Instead, at the time of Epstein's study, NBC relied entirely on its crews in London, Paris, Rome, Tokyo, Saigon, and Hong Kong to cover the whole world. The situation has not changed measurably since, with Teheran (or later San Salvador) replacing Saigon as the "troublespot" assignment. The U.S. picture of the rest of the world is conditioned by hard technical realities such as news crew deployment.[8]

Satellites and other technologies will do less to "democratize" international communication than is often thought. Ideally, the existence of a satellite that is technically capable of indiscriminate communication over the entire globe would result in greater two-way sharing of information, but this will not be the case. The technical complexity of satellite management will probably result in even greater centralization of control in the Northern Hemisphere than has been the case.

An example of the new reality posed by satellites surfaced in the late 1970s when it was revealed that the French government contracted with an American data processing firm to store its top defense secrets. The French government was aghast to learn later that this information was being stored in a data bank in Iowa. The computer/satellite grid employed by the processing firm knew no national boundaries. Defense secrets and auto sales figures are all data, and the master computer routinely sent the French secrets to Iowa in the blink of an eye.

Although the lodging of French defense secrets in the U.S. was inadvertent, much vital information about the rest of the world comes here more intentionally. The U.S. Landsat satellites are highly sophisticated birds capable of forecasting crop failure, identifying mineral deposits, even counting industrial production, all from 22,300 miles in space. They can scan the entire surface of the earth, but U.S. interests control the information. This is another concern of the New International Information Order.

The more active incursion of American telecommunications into other countries is also an issue. This dominance has been a fact for years and would become more prominent under "free flow." The experience of the infant formula controversy of the 1970s was probably a precursor of such situations in which Northern and Western interests used media in Third World countries to directly promote products and ideas.[9] New technologies make such activities more possible.

Advocacy on the NIIO issues has focused in forums beside UNESCO. Another UN-related agency, the International Telecommunications Union (ITU), is responsible for international agreements regarding the allocation of frequencies for communication. Without order in these matters, confusion and jamming of signals would develop and the world's communications could shut down entirely. The ITU also allocates space for satellite communication which has become more scarce as more satellites have been put up.

At the ITU's World Administrative Radio Conference (WARC) in 1979, many Third World countries argued that their interests would be best served through assurance that space be available for their use. "Free flow" proponents argued that countries that have no satellites should yield space to those that do, thus perpetuating dominance of the major telecommunications corporations.

The "free flow" group largely won out and in 1980 the Communications Satellite Corporation announced plans to use scarce space on the equator over the Americas for a series of up to 20 "direct broadcast satellites" for domestic U.S. entertainment use. This in spite of needs that other countries in the hemisphere might have for the scarce space. Further WARC hemispheric conferences on these issues are scheduled throughout the 1980s, and some church groups plan to assist Central and South American countries in organizing active presences at those events.

The NIIO involves the desires of less developed countries to have control over information about them and over their own domestic information services. Their opponents in these matters are not only governments in the North but also transnational corporations, whose power, particularly in the telecommunications area, has grown almost immeasurably.

The MacBride Report puts it this way:

> In the debate on international communication the role played by the transnationals has become crucial. Not only do these conglomerates mobilize capital and technologies and transfer them to the communications market; they also market countless sociocultural consumer goods which serve as a vehicle for an amalgam of ideas, tastes, values and beliefs. The transnationals exert a direct influence on the economic production ap-

paratus of the countries in which they operate, as well as playing a part in commercializing their culture, and can thus modify the sociocultural focus of an entire society.[10]

It has often been said that Americans little appreciate the effect their everyday consumption has on the rest of the world. This has usually referred to consumption of food or energy but applies equally to consumption of information. The U.S. is an information-rich society and the power of U.S. information over the rest of the world continues a pattern which should cause some discomfort.

Section III

Converging Technologies

Introduction

It has been suggested that the United States, and indeed the world, is in the midst of an electronic revolution that will forever transform it. Some scenarios suggest that within the next few years, nearly all essential activities will be done from home by means of electronics. These activities will include some shopping, all banking, all communications (including what is now done by mail), and even work. This new age is being ushered in by a complex interaction of developments—some dating back to the introduction of telegraphy, others to more recent events.

A great part of communication and commerce has been handled electronically for quite some time. Newspapers have used electric "wire" services for nearly a century. Ticker tapes have been a fixture in the financial community for decades. Transportation has used electronics for years. The computer had transformed much business and commercial activity by 1970.

Many of these developments, however, have been largely invisible to the average consumer. Only when consumers begin to have access to some new developments in the home do they see these long heralded changes actually happening. Many people can now bank almost entirely by phone. The telephone company also offers many people services that would not be possible without sophisticated electronics, such as "call-waiting" and "call-forwarding." People who have cable television already are getting news and other information from alphanumeric (letters and numbers) displays on their TV screens. The new age will be characterized, however, by the *convergence* of these and other technology-based opportunities into one or very few "pipelines" that will bring all such services into the home.

To more clearly describe the developments, it is important to look at the service that will make most of this possible, the emergence of cable television. For most people, the emergence of the "home information center" has been and will be made possible by the earlier emergence of cable television, which is "broadband," offering a wide spectrum of channel carrying capacity.

Chapter 8

Cable Television and the Broadband Revolution

Cable television began simply as a means of bringing television to the home over a cable. Most people have always been able to receive their television programming "over the air"—that is, by the *broadcasting* of the television signals over radio waves from local (nearby) television stations. A few people, however, have never been able to receive the signals well, primarily because of distance or physical barriers such as mountains or tall buildings. The FCC's allocation and placement of television stations has needed to strike a balance between providing enough stations to serve most people and keeping the stations' signals at low enough power (wattage) so they do not conflict with each other or with stations in nearby communities. As a result, there are "dark" areas in every station's service area where its signals cannot be received. These are usually in valleys or other low areas near the edge of the station's "signal contour" (its coverage area).

In some areas of the country this problem is more acute because there is a greater population density and thus a large number of people who cannot receive television clearly. These areas tend to be in the more populous but hilly areas of the Appalachian Mountains in the East and South, and the less populous rural areas of the West. Some cities, such as San Francisco, are so hilly that signals are poor within a few miles of the transmitter, but in most cases, poor reception occurs at some distance from the city's center where stations are located.

Cable television began as a way of solving this problem in mountainous rural Pennsylvania. In Lansford, Pennsylvania, some 60 miles from Philadelphia, people had trouble receiving Philadelphia's television signals. In the early 1950s a local television salesman, Robert Tarlton, solved this problem by putting up a large master antenna on a mountain to receive the weak signals and, after boosting their strength with an amplifier, sending them down the mountain into individual homes by means of coaxial cables. The same idea developed in rural Oregon, and cable systems soon emerged all over the country to enhance television signals for viewers in remote areas.

To cable's early entrepreneurs this seemed to be the limit of its application. After all, why would anyone who could receive television signals "free" want to pay for cable service? Cable's acronym reflects this bias. It is called CATV for "community antenna television," and the basic system is just that: It consists of an antenna that receives signals, a "head end" that enhances the signals and puts them on a "trunk cable" that carries the signals down streets to "drops" where cables split off the trunk to individual television sets in homes. The trunks require line amplifiers every 1,500 feet or so to boost the signal, but the operation is basically that simple.

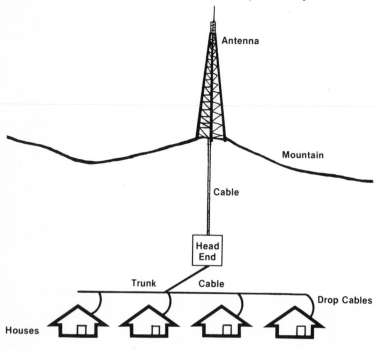

Basic CATV System

All the recent developments of services brought into the home by cable TV are based on this structure. A fundamental question that emerged at the beginning was whether cable constitutes merely a "signal enhancement" service or whether it is a source of signals that potentially could compete with other sources such as conventional television.[1]

Over-the-air broadcasters were immediately suspicious of cable because of its ability to disorganize the market arrangement they depended on. Their worries were primarily over the ability of cable to bring in distant signals that could not otherwise be received in the community. Theoretically it was possible for cable to ruin the plans of a local television station to show a certain program by making that program available in the community by carrying the signal of a station in another community that had the program already.

Cable blurred the boundaries by which television stations had established exclusive rights to program material. Program syndicators and broadcasters had assumed that exclusivity established the value of a given program to advertisers (remember, *they* are the customers). Their thinking was that ephemeral experiences like television programs were valuable only if they were carried exclusively. Doubt has arisen recently over whether this judgment is justified but for the first 20 years of cable's existence it was a major reason for broadcasters' resistance.

The other major complaint broadcasters have about cable is what they call "siphoning." Because it can bring in outside signals a cable could have the effect of "siphoning" audiences away from broadcasters, thus introducing additional competition. The term also implies that cable *could* attract only *certain* audiences, perhaps the cream of the crop, leaving the dregs for broadcasters to sell to their advertisers.

In the early days, the possibility arose that cable might create tremendous competition (as many as three or four additional channels). The more recent development of 100-channel cable was not anticipated by many broadcasters in the early 1950s.

The pattern of allocations of television licenses was set before 1950 and established the situation that still dominates. Most communities (markets) have three very high frequency (VHF) channels—numbers 2 to 13—and one or more ultrahigh frequency (UHF) channels—numbers 14 and above. The basic difference between VHF and UHF for the viewer is that UHF signals are not as good or as easily received as VHF. The FCC's decision to license VHF first, and allow UHF to develop more slowly, left UHF at a disadvantage that has still not been overcome. Many early TV sets had no UHF receiver and thus could not even get the signals. Congress passed a law mandating all TV sets to receive all channels beginning in 1962.

The British had dealt with the UHF parity problem by introducing UHF as the color service, thus linking it to that innovation. In the U.S., UHF continued to flounder due in no small measure to its potential threat to VHF-based television interests that were just establishing themselves in the 1950s.

The FCC's allocation scheme resulted in placing only three VHF stations in most of the major markets (the 25 largest metropolitan areas). This resulted in the evolution of only three television networks, each of which could have one affiliate in each city. Networks were a legacy of radio, where local stations had found a great advantage to contracting with a network from which they received high quality "national" programming. True networks are typified by a contractual arrangement in which the local station agrees to carry a certain percentage of network programming and receives payment in return. Many arrangements are *called* "networks" but are not these contractual relationships that result in the network paying the station.

The network system seemed to make sense for television as it had for radio. In radio, however, more stations had meant more networks. (Four survived into the television age: ABC, NBC, CBS, and Mutual). Television initially had four networks. The fourth, the Du-Mont Network, failed in 1955 largely due to its inability to snare a large enough share of the national audience because of the lack of markets where four VHF networks could compete.

During the early years, the rather complex arrangement governing television use of motion picture and sporting events developed, again with an eye to establishing structures to protect the interests of existing broadcasters. In 1956, motion picture distributors shifted from their early mistrust of television and its threat to the film industry to a more cooperative stance, and hundreds of films from before 1948 were released to television.[2]

To protect their interests, broadcasters arranged to use these films (and later sports) *exclusively* in their markets. According to an FCC staff report, no logical reason for such a thing seemed to exist other than the image of competitive edge, but such provisions, called *syndicated exclusivity rules,* were eventually written into regulations by the FCC. These rules were a complex arrangement that prevented two stations in the same market from carrying the same

film or program within a period of years. Buying the options for programs that were never used by the purchasing stations could
sometimes mean films or shows being "blacked out" for years.

Such a system could easily be disturbed by any innovation that
would allow signals from other markets to enter over a cable. Producers and copyright owners felt threatened; the sale of rights to a
program to be shown in Pittsburgh could mean its automatically being carried in Harrisburg if that city's cable system chose to pick up
that station that day. As long as cable is seen only as a signal
enhancement or reception service, the station in Pittsburgh and the
copyright owner can have no control over where its programming is
carried on the cable.

UHF stations were threatened not only by possible competition
from "distant signals" but also by the possiblity that their signals might
not be carried by a cable, thus putting them at a further disadvantage.

VHF broadcasters saw cable as an interloper, potentially bringing in competitive signals, possibly disturbing the local market structures that had evolved and that supported the three-network system.

The broadcasters naturally moved to the FCC for relief from the
"threat" of cable. In 1964, however, the first time the issue of the
FCC's jurisdiction over cable came before that body, it decided it had
no jurisdiction because cable is a *receiver* (or signal enhancer) service, not an *origination* service. The Commission was to change its
mind and in 1966 declared jurisdiction.[3]

The commission's regulation of cable actually began in 1962
with its Carter Mountain Broadcasting Company decision. That company was restricted from using an FCC-licensed microwave link to
carry imported signals. This decision limited the means by which
signals could be imported. The FCC declared jurisdiction in 1966 by
issuing rules that did not allow for signal importation in the 100
largest markets unless the cable applicant showed that local broadcasters would not be harmed. This had the result of slowing and even
stopping the development of cable in those largest markets.

The logic of cable development is different for smaller and larger
markets. In the small markets, generally ones distant from an abundance of television stations and in towns on the edge of their
markets, the advantage for the consumer of subscribing to cable is
that it is the only way of getting television or getting it clearly. In
many large towns, cable is the only way, for instance, of receiving
public television.

In larger cities, where a greater concentration of the population lives within the good reception area of a number of local stations, cable is an advantage only if it brings in signals and services that could not be had otherwise. For most cable systems this means importing distant signals from an independent, non-network station that programs mostly movies and sports, thus adding to the available sources of programming for the consumer.

Given this logic, the FCC limits on importing distant signals for the larger markets was equivalent to prohibiting cable for them. Only rare instances in the large markets made cable desirable as a "signal enhancer."

Because its rule obviously cooled cable development, the FCC issued a second rule in 1968. This rule replaced the earlier one with a requirement called "retransmission consent": A cable company wishing to import a distant signal was to get the permission of the copyright owner before doing so. This complex plan did not work because of the enmity between broadcasters and cablecasters. Consent was not forthcoming.

Another complex plan devised by the Commission in 1970 allowed distant signals to be carried but required cable companies to put revenues into public broadcasting in exchange for this right, and to substitute local advertisements for the ads that might be carried by distant signals. This rule was also too complex, so other rules were needed.

In 1972 a consensus agreement was reached among the Commission, broadcasters, cablecasters, and copyright owners that established complex guidelines for dealing with distant signals, leapfrogging (importing a signal from a market beyond the market nearest to the cable system), public access channels, and other issues. These rules were to protect the interests of all parties by limiting the chance that distant signals would compete directly with local stations. For instance, in the top 50 markets, the distant signal could not be transmitted if it carried a program for which a station in the local market had exclusive rights. Communication attorney Susan Green writes,

> The practical effect of these rules is to curtail the importation of syndicated programming to the largest markets because the most attractive programs will have exclusive contracts in those markets.[4]

For example, the cable company in San Francisco could not

carry the "Beverly Hillbillies" from its station KWGH in Denver if one of the local San Francisco stations had the "Hillbillies" under contract. The complexity of blacking out only certain programs and the fact that in a market of any size most programs and movies are under contract to someone made these rules restrictive for cable.

Other aspects of the *1972 Rules* that stemmed from the consensus agreement deserve mention. The Commission required cable systems to black out distant network signals during the hours they were being carried by local network affiliates. Thus viewers in Bakersfield could see all three of their local network affiliates and the three Los Angeles network affiliates except during the hours when they were all carrying network shows. Then only the Bakersfield stations could be carried, thus insuring that the local broadcaster benefited from the audience numbers attracted by the network shows.

The *1972 Rules* also mandated that, in the larger markets, cable systems would need to provide "community access" channels that would "originate" local programming to allow opportunity for local self-expression. Access continues to be an important aspect of cable service.

Finally, the *1972 Rules* required all further cable systems to have the capacity of carrying at least 20 channels instead of the 12 that had been the rule until then. This was a big step for the cable industry because it implied the need for additional hardware in the home, hardware that was one of the first harbingers of the new information age. Conventional 12-channel cable systems could hook directly to the back of the home receiver and channels could be selected by using the tuner (channel changer) on the set. When a system of more than 12 channels is installed, the cable company must also provide a *converter,* a box containing a tuner with more channels that is hooked to the TV set.

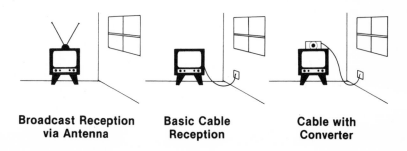

Broadcast Reception   Basic Cable   Cable with
via Antenna   Reception   Converter

As is the case with most FCC rules, cable systems in place or planned at the time of their introduction did not need to comply with the rules. Thus, in 1980, eight years after the *1972 Rules,* 80 percent of American cable systems were still only 12-channel systems without public access channels.

Cable has potential beyond access, origination, and local and distant television station carriage, however. Cable can carry one or more subscription (pay cable) channels for which subscribers pay an extra fee to receive special services on either a per channel or per program basis. This ability could be a great boon to special audiences who are not large enough to justify a network program, but to whom a program could be beamed on a pay-for-play basis, and to entrepreneurs who could profit from providing special movie or sports channels for which subscribers would pay an extra charge.

Pay services on cable were initially discouraged by the FCC with 1975 rules curtailing the access such channels could have to films and sports. Those rules were struck down by the courts in the *Teleprompter* decision in 1977 and the FCC has since relaxed them further so that pay cable has begun to flourish.

The first and best known pay cable service is Home Box Office which went on the air in 1975 using a satellite to distribute its channel of movies and sports to individual cable companies.[5] The system works in this way: A local cable company contracts with HBO to offer HBO as a subscription service. The cable company then puts up a satellite receiving dish pointed at the satellite HBO uses (an RCA Satcom satellite) and charges interested customers an additional fee for HBO. It divides the income with HBO, keeping some of the fee but passing most of it along. In 1980, the average basic cable fee was $8 per month and the average fee for subscription service such as HBO was an additional $8.50.

For their money, subscribers get the HBO channel which carries first-run motion pictures, sports, and entertainment programs, all without commercial interruption.

By 1980 pay cable had become the driving force behind cable in major markets with a number of companies joining HBO as premium services. Showtime was HBO's nearest competition, with Prism running third in number of subscribers. NBC and CBS both were planning to enter pay cable with cultural programming. HBO planned a second movie channel. Warner Communications offered The Movie Channel. Bravo and Escapade, a combination service of R-rated movies and cultural programming, was another possible channel. In addition, a stunned PBS audience learned in January 1981 that BBC had agreed to sell its programming first to a pay channel owned by Rockefeller Center beginning in 1982. Thus a major source of attractive entertainment programming on PBS was shifting to pay cable.

Given all the services, it was theoretically possible in 1980 for a subscriber to a large cable system to spend $80 a month on cable programming. Fueled by pay cable, cable development between 1978 and 1980 resulted in a near doubling of industry revenues from $800 million in 1978 to $1.5 billion in 1980 of which $200 million was for pay cable.[6]

Pay cable is merely the "premium" service cable can bring over and above local stations. The 1970s saw the development of a peculiar kind of distant signal known as the "superstation." This idea, pioneered by Ted Turner in Atlanta with his WTBS-TV, is technically no different from the standard carriage of a distant signal by a cable system. The unique aspect is that Turner actively encouraged as many systems as possible nationwide to carry his station. Other "distant signal" stations are often picked up and carried by a few relatively nearby stations.

Turner pioneered the idea that an independent station might be programmed with popular services, sports, and movies to make it attractive to cable systems as an alternative channel. Turner furthered his cause by purchasing a satellite uplink (an antenna used to send signals *to* the satellite and then down to earth again) and leasing space on the satellite, thus offering WTBS signals to any cable systems who wished to pick them up. He planned to make his money by selling advertising on the basis of the much larger audience that viewed his station on cable. It was not clear initially whether this scheme would work, but hundreds of cable systems do include WTBS in their service.

There are other "superstations" besides Turner's. WGN in Chicago and its sister, KWGN in Denver, and KTVU in Oakland, California, are all superstations. They differ from Turner's WTBS in that they have achieved that status unwillingly. WGN sees its role only as a local independent VHF station in the country's third largest market. It carries sports, movies, and specialty series. Another company, called a "resale common carrier," receives WGN's signal, without its consent, on an antenna in the Chicago area, and sends it via an uplink and satellite to cable systems all over the country. The arrangement is perfectly legal, but is troubling to WGN.

With the local signals, distant signals, superstations, and pay channels carried by cable companies, a proliferation of services would seem to be available to the homes that have cable (20 percent of the TV households in 1980). Does this mean that cable will provide a greater diversity because of the greater number of channels available? With most cable systems having only 12 channels, and those additional signals available carrying primarily movies, sports, and reruns (even the vaunted HBO carries old sitcoms during part of its day), it is hard to call cable revolutionary or diverse. This proliferation will not necessarily mean diversity just because it provides greater numbers of channels.

Access and origination

Cable television does appear to be different in its provision of *origination* and *access* programming. Under the *1972 Rules,* cable was brought under the concept of localism and public trusteeship by being required to provide a certain amount of local origination programming—programming it produced on cable only. It was also required to provide studio facilities and channel capacity so that citizens could produce original programming. These provisions did not apply to the earlier markets, only those proposed after 1972.

Because cable companies must get city approval to install their cables over city streets, as well as FCC approval, they must go through a local franchising process. This local franchise process has often been more lax than the FCC's, but in some cases it has been more stringent, especially about access.

The best examples of this are New York City's two franchises for Manhattan, both of which must provide access channels on a "noninterference," "nondiscriminatory" basis. These access channels are widely known for carrying some of the worst television ever, including nudity and sexually oriented talk shows. At the same time they have resulted in true diversity, including some wonderfully spontaneous artistic expressions, and some of the best community service programming ever seen. Jerome Alderson reviewed the Manhattan access programs for *Columbia   Journalism Review:*

> Taken together, the access shows define a new television esthetic—an esthetic of reality, as opposed to the network esthetic of illusion. On access, people appear more natural, talk more freely, and address more issues of concern than on any other form of television. To this may be added the spontaneity of the live transmissions preferred by many access producers. Access is the progenitor of programs like "Saturday Night Live" and "Real Peo-

ple," but unlike the subjects of "Real People," the real people on access don't need Hollywood types to introduce them and thus distance their own real lives from the real lives of their audiences.[7]

In other communities, access channels are used by churches, schools, community groups, and colleges as forums for local, civic, education, and religious affairs. Councils of churches in Cincinnati, Pittsburgh, and other cities regularly produce access programming of high quality. Local congregations in communities with access are able to produce programs on a one-time or regular basis, often at no charge and with the assistance of cable company technicians.

Origination programming, in which the cable company produces local programs on community or public affairs, is also a significant alternative to commercial television. Given the relative abundance of channels available (now shrinking under the pressure of pay services and distant signals), cable companies can produce true alternative programming. The cable company in Audubon, New Jersey, for instance, provides live origination coverage of local city council and county government meetings. While not always exciting, this sort of programming does serve democratic purposes. The cable system in McPherson, Kansas, produces a local news program using journalism students from a local college—a good experience for the students and an additional local news source for a small city.

Access and *origination* are threatened, however, by the demand for more channels for pay and entertainment services. In 1976, Midwest Video, a small multiple system operator (MSO) took the FCC to court in Chicago over the FCC's rules mandating access. The Chicago court, which has always tended to be more open to industry-originated complaints than other circuits, ruled that the FCC has no jurisdiction to require access. This ruling has not yet had the effect Midwest Video intended, that of freeing up access channels for commercial or pay services. Instead, many cities have taken a cue from the FCC and New York City to require access as part of their own local franchising process. This local authority based access is often better than the FCC required, with recent franchises in some major cities offering four to six access and origination channels and full studio facilities exclusively for their use.

The horizon, however, does not look bright for access and origination. The cable industry still looks on them as impediments to their market potential. In 1980, the NCTA (National Community Antenna Television Association—cable's primary trade group) had a bill introduced that would have prohibited cities from requiring access. This bill stalled but the conservative Congress elected that year appeared likely to reintroduce such legislation.

Even if Congress does not outlaw access it is likely to fade away. It is underused in many communities, and cable operators are warning the public that they will make the case for access being dropped when their franchises come up for renewal. In many cases, cable operators can point legitimately to lack of use and public interest. In other cases, they will probably underestimate the importance of access and origination in order to justify getting rid of them. In either case strong public support and interest will be important if access is to continue. The likelihood of access being dropped will be particularly great in medium-sized and small markets where public awareness of access is lower and where there is less of the big city tradition of requiring it.

"Two-Way" Cable

The cumulative effect of overturning the *1972 Rules,* the Midwest Video decision, and the FCC's 1979 move to deregulate cable has been to remove almost all the impediments to cable's development in the larger markets discussed earlier. At the beginning of 1980, very few major cities were wired for cable. Only New York, Atlanta, and portions of Los Angeles had working cable franchises at that time, but during that year Minneapolis, Pittsburgh, and New Orleans let franchises. Philadelphia and Chicago faced the possibility of not having cable systems in place until 1990.

The major cities are important to cable because the majority of the population lives in them. Cable probably will not achieve the economic power to realize its potential until 35 to 40 percent of all homes are hooked up.

Cable's primary advantage over broadcast television is its ability to deliver a greater number of signals than can be put on the air. The radio-TV spectrum is limited and must be regulated carefully to keep stations from interfering with one another. The physics of the coaxial cable used in cable allow it to organize a great number of television and radio signals so that with little or no interference 30 or so of them can be delivered over a single cable. Paired or bunched cables can deliver more signals to home "drops." The communications world was surprised in 1980 to learn that the franchise for cable in the borough of Queens in New York City would have four bunched cables and deliver 105 channels. The more common pattern in the early '80s has been for 30 to 54 channels to be offered.

Cable differs from broadcast in another important way: its ability to carry signals in two directions. Theoretically, the local subscriber can "interact" with the cable "headend"—something never possible with conventional TV. Most older cable systems are not "two-way" because, as the cables were strung, proper provisions were not made at the line amplifiers along the way for signals traveling upstream to get by.

Newer systems have provisions for "two-way" flow with taps strung around the amplifier allowing signals to get by much as salmon can swim up "ladders" around dams to get upstream to spawn. Most cable systems have two-way cable in place but are not two-way simply because they do not have equipment in the home which allows the viewer to "interact" (send messages back to the headend). They also do not have the necessary technology at the headend to receive those interactive communications. Those systems that are truly interactive, that do have the capacity to allow viewers to talk back are not fully "two-way" because the viewer can usually , respond only by pushing buttons. The signal coming "downstream" is full video, and the viewer can only tap out yes or no responses. It would probably be more precise to call these systems "one-and-one-quarter-way interactive" instead of "two-way" but cable industry and government representatives insist on the "two-way" designation.

The capability of a cable network to be two-way lies not only in the physics of its cables but in its configuration—either tree structured or star structured, switched.

The tree structured system has a "headend" from which signals originate and "drops" or homes ("tailends") where viewers receive the signals. A two-way system allows viewers to send messages back "upstream" but because of the cost of having viewer video equipment in all homes and the complexity of deciding who is to get air time and how, the few responses are limited to simple yes or no responses, or coded messages that are read and recorded by a computer. A star structured, switched system is different because there is really no "headend" and no "downstream" or "upstream" direction. Instead, receivers at home communicate with one another through a switching center. Thus the star structured, switched system is truly two-way interactive in a way that the tree structured system can never be.

It may be obvious that the star structured, switched system is already the most common pattern in the United States. The telephone system is star structured, switched. The star structured, switched system would not be able to handle all the traffic that would demand time and be anything other than a "video phone" system. Some organization is necessary and it is through this organization and control that the cable company derives its income. Either by charging subscribers or by selling advertising time, cable companies parlay their control over the network into income.

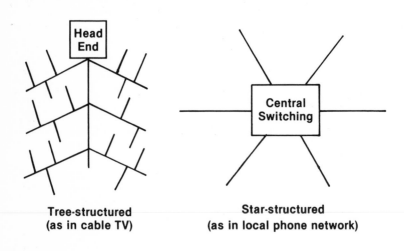

Tree-structured
(as in cable TV)

Star-structured
(as in local phone network)

Tree-structured and Star-structured
Switched Networks

For the future, it is important to notice that the two configurations are not physically incompatible. That is, it is theoretically possible for the same cable network to operate as a tree structured one and a star structured, switched one simultaneously. The signals would need to be organized and switched accordingly. This is important because it is possible for the phone system and cable systems to use the same wires, poles, drops, and home terminals at some point in the future.

A Sophisticated System
During the period of experimentation that followed the Midwest

Video and Teleprompter decisions in the late 1970s, Warner Cable introduced a sophisticated, experimental two-way system called QUBE in a suburb of Columbus, Ohio.

A basic description of QUBE (it is not an acronym) reveals some helpful ideas about the technological capabilities of cable. QUBE is a tree structured system employing sophisticated computers at its headend *and* in each home. The lowly cable converter box in QUBE is replaced by an "addressable" home terminal through which the viewer can respond to questions on the TV screen and to the products that are advertised, or order special pay programs on a per program basis. "Addressable" means that instead of being constantly hooked up to the main computer, the home computer constantly is "asked" by the main computer to report on the positions of its buttons. On receiving a response, the main computer at the headend adjusts that terminal accordingly, changing channels as requested or recording responses. The process is so fast and the "address" scan so frequent that a viewer switching channels would scarcely notice the time between a button being pushed and an order being executed by the computer.

QUBE offers subscribers 30 channels of programming in three tiers (levels) of service. For their basic subscription ($10.50 a month in 1978) subscribers get the basic tier of local television programs and some of the snazziest access and origination programming in cable. For the "premium" tiers, viewers can request specific programs on a per play basis, including adult films, sporting events, and first-run movies. Viewers can also receive sophisticated security services over the cable, including fire, medical, and burglary/emergency alarms.

QUBE subscribers claim to be most intrigued by their interactions with the system's buttons through which they can register opinions about questions asked on the screen. This feature allows a talk show host, for instance, to ask viewers whether they agree with a guest. The responses are instantaneously tabulated by the central computer and reported on the screen for all to see. This system of buttons is also used by QUBE's educational channels to take roll and administer true/false or multiple choice tests in courses taken for credit. The computer grades the tests by reading the response buttons of enrolled viewers.

The interactive feature is also popular with market research firms who have long used Columbus as a "test market"—a city so representative of the populations manufacturers want that products can be tested there and changes made before the products' national release. These firms can use QUBE subscribers for questions about products and packaging, getting responses instantly from the QUBE computer. In an early test, *US* magazine used QUBE to decide which faces on its cover would sell the most magazines. The 1978 test revealed that John Wayne and the Incredible Hulk would sell.[8] Another feature of QUBE that may come to be a staple of cable in the future is the "infomercial"—a program that is a paid commercial for a product or service. During QUBE infomercials, viewers can place orders by pushing buttons. Since the FCC has no rules governing cable advertising, such a program (which would be prohibited as a "program length commercial" on broadcast television) can be carried and need not be identified as a commercial message. The QUBE audience is a prime one for such treatment, not because it is large but because it is a generally "upscale" audience with a lot of disposable income.

The demographics of the cable audience generally make it an attractive target for sponsors, and subscribers to premium services such as QUBE are just the sort of people certain advertisers would like to reach. It therefore seems highly unlikely that premium services such as HBO, Showtime, and others will continue to be commercial-free. Indeed, *Broadcasting* magazine announced in a 1980 article that "the question is not whether there will be advertising on cable, but when. . . . "[9]

In spite of the upscale audience, QUBE loses money. It was intended to. American Express and Warner, the owners of QUBE, designed it specifically to test sophisticated ideas that may not be used in actual working cable systems. The test is going well, and many of QUBE's innovations, particularly its marketing features and use of electronic funds transfers (EFT) for billing will probably find their way into many more cable systems.[10] Its considerable access and origination programming are probable money losers and may not be replicated elsewhere.

QUBE provides a good lesson in the potential benefits of future cable systems, but it also provides a look at one of the greatest problems people face as they become subscribers to such a system—the problem of protecting their privacy.

The protection of privacy becomes a worry when it is noted that QUBE subscribers can empower the central computer to purchase products for them, deliver books and films to their homes, all through electronic funds transfers using subscriber bank accounts and charge card numbers. Warner officials are matter of fact about the issue. John Wicklein, an authority on the subject, quotes a QUBE official:

> People will simply have to accept that they give up a bit of privacy for it . . . Beyond that, we'll try to protect their privacy all we can.[11]

According to Wicklein, QUBE officials admit that even respondents to multiple choice questions can be and often are listed by the central computer. The computer also keeps up-to-date data on subscribers' charge accounts and buying habits as purchases are made. In the eventual case that telephone and cable are delivered through the same network, all phone calls made and received by a subscriber would be listed as well.

Wicklein paints a simple scenario in which QUBE clients could request specific damaging information about certain subscribers. Although such a system would not collect information on people that might not be otherwise available, it would be the first time all such data would be available in one central file all the time. In addition, cable companies seem interested in reserving the right to collect such information about their subscribers. At least it is unlikely that they would support legislation to limit their ability to do so. They seem to want subscribers to trust them. Those who remember how the telephone company dealt with "confidential" phone records during the 1960s may not be comforted.

Cable is Big Business

The late '70s also introduced cable as big business. The FCC announced in 1980 that cable's revenues had increased by 20.3 percent in 1979 to more than $1.8 billion. Pretax profits had increased in that one year by 45.4 percent to $199.3 million.[12] The *Philadelphia Inquirer* quoted Brother Richard Emenecker, Pittsburgh's municipal director of cable television operations, on the future of cable as a corporate enterprise:

> The industry hasn't sorted itself out yet, and it won't probably for ten years. At the end of ten years the cable wars of the 1980s will be over, and all of the cable systems will be owned by ten large companies.

The *Inquirer* writer continued:

In the beginning, cable television was in the hands of many small "mom and pop" companies. But now, the corporate giants are taking over. In the last two years, Storer Broadcasting Company, a large Missouri-based firm that owns radio and television stations, has acquired 47 cable franchises in New Jersey. And The New York Times Company recently signed an agreement to buy 55 cable franchises in South Jersey.[13]

Since the FCC has pulled out of cable regulation, such moves are not monitored by anyone, least of all the small city councils that let the original franchises. The "cable wars" Emenecker referred to are knock-down, drag-out controversies between large and small corporations for the franchises for large and small markets, and the forums for resolution of competing claims are the local governments.

Cable television means big money for entrepreneurs and many cities want a piece of the action. More than a little graft is involved. In a famous case, the chairman of Teleprompter Corporation, the largest cable company in the country, received a five-year jail term for attempting to bribe officials in Johnstown, Pennsylvania, to preserve Teleprompter's franchise there. The same man, Irving Kahn, bounced back from prison and organized the small system of 55 cable companies that was sold in 1980 to The New York Times for a reported $100 million.

A case of corruption in a large market occurred in 1978-79 during the franchise process in Houston. A reporter for Texas Monthly reported it this way:

> When the Houston City Council awarded cable television franchises last winter the decision had nothing to do with the quality of service the applicants could deliver, but everything to do with the power and influence they had behind them. The city handed the franchises over to political and business insiders. Then some of the lucky insiders cashed in their chips by reselling their franchises for millions of dollars. No one in City Hall seemed to mind.[14]

The practice of setting up dummy print corporations that can win the franchise because their ownership is local or has clout and then in turn sell the franchise to someone else is not illegal unless it violates stipulations of the city's franchise. It is estimated that Warner Cable paid $30 million for the Houston-Gulf Coast franchise in the above incident. A much more common franchise strategy involves "rent a citizen": An out-of-town franchise applicant enters into partnership with a local school or community group, giving it a share of interest in the franchise in exchange for their local "good name." An example is the 1980 alliance between a Canadian cable company and the University of Pennsylvania which applied for a franchise in northeast Philadelphia.

Obviously, cable television and its regulation began the '80s in an uproar. Things should settle down. The telecommunications market will organize cable into a well-oiled component of the information society. The next chapter will examine how that will occur and how cable will fit into the information age.

Chapter 9

The Future in the Home

In the late 1970s sociologist Daniel Bell first used the term "information age" to describe the transformation of Western societies from *industrial* societies into *information* societies. In this transformation, activity shifts from the production of hard goods to the production and management of information.

The significance of this projection may be lost on most Americans. Just as the average person was probably not overly affected by so monumental a development as movable type in its early years, many people may look on the dawn of the information age with some skepticism. Others will miss it entirely because they will not know what they are looking for.

Some aspects of the information age were well in place by 1980, others only began to emerge then. The technologies that will bring the information age into most people's homes were for the most part only on the drawing boards before 1970, but by 1985 many people will experience them in their everyday lives at work and at home.

The development of the information age will be characterized by the convergence of these technologies into a system called *compunications* or *informatics*. Communication processes and technologies and data processing techniques have gradually come to be so similar that the combination of communication and data processing functions in the home seems as inevitable as it had proven to be in commerce before 1980.

By the beginning of the '80s, small, sophisticated "home computers" had become the hottest thing on the market with sales volume far outstripping projections. Radio Shack, Texas Instruments, and Apple had come to be household words as increasing numbers of upscale families added a home computer to their inventory of appliances. Radio Shack, which pioneered the market, advertised its computer as a source of family fun, a learning aid for children, and a way to ease the burdens of household bookkeeping.

Much of this development in computers was fueled by the integrated circuit (IC) computer chip. These microprocessors, virtually unknown before 1970, have revolutionized many parts of the information and data processing field. These technologies developed rapidly and quickly dropped in price so that small, handheld calculators were available by 1979 for less than $30 and did the same work as multimillion dollar computers of the 1960s.

It is important to become conversant with the technologies and developments that are converging to create this new age. The first new nonbroadcast technology was cable television. From its lowly beginnings as a signal enhancement service, cable will become the primary conduit of telematics or informatics into the home, a sort of multilane highway over which entertainment, information, data, banking, and a myriad of other forms of intelligence will travel in both directions.

Cable is the "Trojan horse" of the new age: It has entered many homes and will enter many more in coming years as an entertainment service, but it carries the ability to become something far different. Just as the telephone has accustomed users to the process of instantaneous communication, so cable will gradually condition its users to rapid information interface. Already, QUBE and other systems have packages to train subscribers in their use.

Having experience with the new technology is crucial. One man who began banking by phone was at first openly suspicious and hostile to the process. His attitude shifted to acceptance and finally to actual pleasure at the speed and efficiency with which he could bank and pay his bills. Never fond of banks, he has enjoyed the convenience of entering his bank only once in 18 months.

Nearly all the current change in compunications/informatics/telematics has been fueled by recent, rapid developments in communication regulation. Although the FCC can regulate broadcasting because it operates on a public resource—the airwaves—such a mandate is less clear with new technologies. This became obvious in the regulation of cable as first the courts, then the Commission itself, considered cable something separate from broadcasting and freed it of obligations, primarily those with a "public interest" focus.

The Commission also handles common carrier regulation, allowing monopoly carriers to operate with no competition in exchange for regulation of their activities and rates. This regulation came about not because of scarcity of capacity but because it was felt that "natural monopolies" existed in areas such as telegraph and telephone, where it made little sense to have more than one phone company in an area. By having one company hook up everyone in a community, the problem of customers not being able to talk to one another because they were on different phone companies was eliminated. Without competition to insure service and keep cost down, regulation had to be imposed.

More than 800 telephone companies exist in the United States, each with a monopoly over its service area. American Telephone & Telegraph Co. (AT&T), the largest corporation in the world, is many times larger than any of the other phone companies. Bell is so large that it constantly has been accused of anticompetitive practices with its few competitors, including a 25-year battle with smaller companies that wish to manufacture telephones or accessories for telephones. In the heyday of Bell sovereignty over phone service, it was even illegal for subscribers to hook shoulder pads to their receivers unless they were installed by Bell.

One piece of customer equipment that brought the house down on Bell was manufactured by Hush-a-Phone Corporation. It was a simple plastic shield that fit over the mouthpiece so extraneous room noise would not interfere with a phone call. Hush-a-Phone and Bell went to court (the cost of litigation eventually put Hush-a-Phone out of business) and Bell lost. It lost again on the question of acoustic couplers (Could customers buy cradles for phone receivers that amplified calls or interfaced with other equipment?) and eventually lost the right to require customers to buy their phones only from Bell. Today, with rare exceptions, customers can purchase any telephone they wish anywhere they wish, often at substantial savings.

Bell and the other companies retained their monopolies in local exchange (or intraexchange) service throughout the period of regulatory upheaval in the late 1970s, but they lost their monopolies on interexchange or long distance service to competing carriers (initially called "private line carriers") beginning in the mid-'70s. The best known of these private line carriers, Microwave Communications, Inc. (MCI), won an important court battle that allowed it to enter the field of interexchange service. By 1981, MCI had signed up hundreds of subscribers in high density areas all over the country who reported substantial savings on their long distance bills by using MCI services.

Because local exchange is regulated as a monopoly, MCI

customers still use conventional phone service locally. When they wish to make a long distance call, however, they bypass those long lines by dialing a special access code before the number they wish to call. The switching computer automatically sends their call over MCI's lines instead of Bell's. At the other end the call is switched back into the local exchange and rings as a normal phone call.

As with most of the coming changes, MCI and other competitive carriers would not be possible without the computer. These computers are a mixed blessing for Bell. It is this reliance on computers and telephone switching that has brought Bell this new competition and also makes it likely that Bell will be a major force in future nontelephone compunication or informatics.

As the number of persons served by Bell has increased, and the complexity of switching (the process of connecting calls) has increased, Bell has made a transformation in switching technology called "packet switching," which is really "computer switching." In traditional switching, callers were connected through a switchboard by hand at a central toll office (remember the phone operators in front of the bank of cables and sockets?). More recently, cables and sockets and switches were replaced by mechanical switches that took the role of the operators. (Most people became aware of this change when their old phones were replaced by dial phones.)

With the advent of the computer it became clear to Bell engineers that if telephone messages could be transformed into data that could be handled by sophisticated computers, greater efficiency and economy could be realized. To understand why, it is necessary to think about the difference between conventional telephone signals and data (called "digital") signals. A computer is merely an extremely complex set of switches that can be either on or off. Series of these on-off signals form the computer's binary (on/off) code of communication. Naturally, the computer is best at dealing with numbers because its code is numeric. Sets of its codes, however, can be used to transmit letters and other sorts of data or energy as well.

A conventional telephone or other audio signal is in a form called *analog*. This is a complex concept, but it helps to think of its meaning this way: The signal that carries telephone calls (voices and other sounds) across old copper wires is "analogous" to the speaker's voice. The signal's intensity increases as the voice does, and it carries dead air when no voice is being carried. New "digital" telephone transmission replaces the system with one in which the voice is transformed into digits instead of electrical impulses on a copper wire, and is sent across the phone network as data, along with other computer-coded data "bits." The phone company's packet switching computers take a phone call and send it across the network to its destination as thousands of digits of information. At the other end it is retransformed into the voice and delivered to its destination as a phone call, all instantaneously. At both the sending end and the receiving end, the call is analog, traveling across the same copper wires as always. But once it reaches the switching office in the sending city, a new computer takes over.

A fascinating thing about the process is the fact that packet switching allows a phone call to be sent in "packets" so that part of it may be routed through Chicago, another part through New Orleans, another over satellite to Denver. But because all computers are on the same network at the same time, it is automatically regathered at the proper destination. This feature allows Bell to maximize use of its lines because there is no dead air. In the old system, any time the callers paused in their conversation, the "open line" to the other end carried just dead air. With packet switching, the dead air occurs only for the callers. On the lines between them, that air is filled with data going somewhere else.

In testimony before the Senate on possible revisions of the Communications Act in 1979, a Bell official disclosed that the packet switching network was well underway. Bell, he reported, was installing large new computers on its network—computers larger than the largest "main frame" computers sold by IBM to businesses—at the rate of *one per day*. This computerization will continue until Bell has transformed its network into the world's largest computer/data/communications grid. This network of thousands of computers connected constantly by Bell's lines becomes one super computer with unlimited capacity. Since these computers do not discriminate between data and telephone calls, they could be used as a data processing grid as well; Bell would like to enter the data processing market with this resource.

Bell's entry there is specifically prohibited, however, by FCC regulations and by court actions. Bell's sheer size and its monopoly position have been problems for years. Bell's ability to cross-subsidize the various services it provides and its power in the market have led to hundreds of lawsuits, including one settled by the Federal courts in 1956. The outcome of that case was a consent decree (a negotiated court settlement) that still holds Bell back from full entry into non-communication markets.[1] Thus even though Bell wants to enter the data processing field with its new computer capacity, it is prohibited from doing so by law.

The convergence of technologies has created a problem with that law, however. The consent decree prohibits Bell from entering noncommunication services. But when a packet switching computer is used, is that data processing, or is that communication? The line is blurred between the two. A further complication is the inefficiency of letting Bell's data capacity go unused. This is the argument Bell has been using to promote its cause for freedom from restraint before Congress and the FCC. Should it be persuasive at some time, Bell would immediately become the world's largest data processing firm and would probably overwhelm many smaller companies such as ITT and IBM. Bell's potential competitors are terrified of such a development and are the primary opposition to any deregulatory moves by Congress or the Commission.

The FCC started to write a new chapter in this area in April 1980 when it essentially deregulated Bell in the data processing field. In a series of new regulations, the FCC allowed Bell to enter "enhanced services" fields using its data processing capabilities. Although this would ultimately work out to be full entry into the data processing market, such a move is in the future because Bell still has to consider the stipulations of the 1956 consent decree in the Federal courts. If Bell does move in this area, it probably will not be before 1984 or '85.

One other technological development by Bell that will shape the development of the Information Age will be Bell's introduction of fiber optic transmission lines. Technology has gradually increased the capacity of Bell's long distance lines to the point that thousands of calls can be handled simultaneously. Fiber optic cable will bring this capacity directly to the home, replacing the old copper phone line with a cable that possesses a capacity several times greater than cable television's coaxial cables.

Once most homes are hooked up to this capacity, the Information Age will have passed its last frontier by coming directly to the home with its multitude of services.

One can envision a system based on Bell's fiber optic cables and computers that would make it possible for every home to have access to hundreds of video channels plus hundreds of information channels of alphanumerics (letters and numbers on the screen) plus two-way interaction with banking, shopping, health, and education services. Even the ability to work at home would be possible, a capability that might be available first to disabled persons.

In 1975, a prestigious think tank, the A.D. Little Company, projected that by 1990, Congress will have made Bell *the* monopoly provider of such services. By that time, according to this projection, Bell will have so completely overwhelmed its competition in data communications and cable television (which it will have been free to enter in 1980), that efficiency will have dictated it to be in the national interest for Bell to become a kind of national utility. This projection may not happen, but it should help dispel the myth that this proliferation of services means a decentralization of communication control. Quite the opposite seems to be true.[2]

Background information on Bell may seem unrelated to a discussion of new services in the home, but anyone who has a pushbutton telephone already has access to the data processing capacity of the Bell network. Already, many people use their phone key pads to bank from their homes. The other communications developments of the age of convergence are further off for the average home, but some people have access to them already. A discussion of each of them follows in the order in which they may become available. Nearly all of them presuppose the development of the two primary services, cable television and packet switching.

Subscription television

Subscription television (STV) has been around a long time. It is basically a conventional television station that sends a scrambled (encripted) signal that must be unscrambled (decoded) by a device rented from the STV station. Income is thus guaranteed through subscriptions controlled by rental of the devices. It is a messy and somewhat complicated process to manage but a number of them have been on the air since 1960.

Broadcasters naturally opposed STV because of its potential to siphon audience and succeeded in getting at least one state (California) to outlaw STV initially. In the late 1970s, STVs became more prominent and profitable and a feud erupted over whether nonsubscribers were breaking the law if they made their own descrambling devices. This was in question because under the law, the airways belong to the people and cannot be controlled. The case was in court for years and, before its resolution, will probably become moot by the development of other scrambling techniques. STV has operated under different sets of FCC regulations much as cable has. Included are regulations that limit the numbers of hours they may be on the air and their ability to sell advertising. The advertising stipulation will probably fall first, as both

STV and subscription cable will turn to advertising as one of their greatest potential sources of revenue.

Subscription cable

This is a far more common service than STV and is available in some form over nearly every cable system in the United States. The early purveyors of cable believed additional channels to be one of the most attractive features of this new service, but such extra services were long in coming because of opposition from the FCC and broadcasters. Initially, cable companies could not even provide additional TV station signals without FCC approval. In 1975 HBO took the FCC to court to win the right to go on cable systems not as another TV station but as a subscription movie service. HBO won the case. Within three years it was carried by hundreds of cable systems and had become a household word. Other subscription-only services sprang up quickly including other movie channels, sports channels, news channels, and a channel that carries gavel-to-gavel coverage of the United States House of Representatives.

These subscription channels and other channels that are not primarily television stations or subscription-supported (such as the three "Christian" channels available by 1981) are possible because of the development of the communication satellite. Cable companies wishing to receive and offer these channels need only construct satellite antennas, and most have done so. Not all extra channels are subscription-based in all cable systems. Some are offered as part of the basic fee as an inducement to basic subscribers. (Major ones such as HBO are always "premium"—available only at extra cost.)

The future for the subscription channels will include advertising. It is inconceivable that the opportunity to offer advertisers access to the "upscale" audience that can afford these premium services will be passed by. To accomplish this shift, HBO and other services will ask their subscribers if they wish to have their fees lowered or not raised in exchange for certain advertisements strategically placed in programming. Market research indicates that a large percentage of subscribers would be willing to have a little advertising in exchange for lower subscription fees.

Satellite transmission

Until direct broadcast by satellite becomes a reality most homes will not have direct access to satellite channels. Instead, satellites are used exclusively for the transmission and distribution of television and other communication material between cities, to cable companies, and by PBS stations. Broadcast networks (CBS, NBC, and ABC) had always been logical because the local stations could depend on them for regular feeds of programming that cost the stations nothing to obtain. The network bore the cost of transmission by phone lines.

When satellite technology developed birds (satellites) that could stay in stationary orbit and act as transmission lines in the sky, and as satellite space became cheaper and more flexible than Bell's land lines, the potential for other networks developed. (Other networks always had been technically possible. Limiting the number of stations per community also had the effect of limiting the numbers of commercial networks to three.) More important, signals could be sent up to the satellite and back down to anyone who had the money to build a downlink (antenna), thus bypassing traditional networking and changing the costs involved. The provider of the signal has the same cost for satellite time whether none or a thousand cable systems pick it up.[3] An immediate result of satellite development was subscription cable, but another development has been the superstation.

Superstations

An additional cable channel available on most systems is the superstation. This is an independent station whose signal is available nationwide by satellite transmission. Superstations are independent stations that program movies and sports and that make extra advertising income by being available to a national audience over cable. The best known superstation is Ted Turner's WTBS in Atlanta.

Turner also owns interests in the Altanta baseball, hockey, football, and basketball franchises as well as the rights to more than 2,500 old movies. In 1979, WTBS began sending its signal by satellite to any cable companies that wanted to use it as an extra channel (attractive because of its additional source of movies and sports) for five cents per subscriber, which covered transmission costs. WTBS's income for the enterprise came from advertising revenue. Two other superstations, WGN in Chicago and KTVU in Oakland, are unwilling superstations; their signals are picked up and boosted to the satellite by third parties over which they have no control.[4]

Turner has built on his experience with WTBS by starting the first 24-hour TV news channel, Cable New Network. This premium service is available on most cable companies around the country. It is 24 hours of news and public affairs programming. Many are skeptical of how long this expensive service can stay on the air. It is commercially sponsored, but must show a sizeable audience to stay in business.

Another news service that has capitalized on satellite transmission has been WPIX-TV's Independent Network News. CBS, NBC, and ABC have dominated national TV news because they could feed current programming to their affiliates. With satellite capability, WPIX in New York is able to provide a "fourth network" news program and feed it to independent stations around the country.[5]

Low power television

In 1980, the FCC stirred up a potential controversy by authorizing a number of so-called low power television stations. These stations would fit into the existing system of TV stations and signal contours and allow for as many as one thousand additional stations nationwide. Because they would operate at lower power than conventional stations (a few thousand watts instead of the conventional stations' 50,000), they would be established without interfering with other stations. As explained by the Commission, the system would usher in a television service resembling current radio service with a number of stations of different power offering a greater variety of programming. Some low power stations would have signal areas of only a few square blocks in major cities or a few square miles in rural areas, others would be slightly more powerful.

This development is viewed with skepticism by many. For a number of years the FCC has considered authorizing additional high power television stations, which would be technically possible. This proposal for what are called "drop-ins" has been roundly opposed by the established stations and by the networks because the stations could theoretically form a fourth network and thus increase competition.

The recent low power proposal is less threatening, however, because no one station could be a significant threat to regular stations' audience levels. Taken as a whole, stations might prove a threat, so conventional broadcasters were mixed in their reactions to the low power proposal. At least one network announced intentions to apply for over a hundred of these licenses, as did Sears, Roebuck and Co., until the FCC put a cap on the number available to one owner.

The FCC touted low power as a way for minority and public interest groups to have access to air time, and some of these groups applied for licenses. The Southern Baptist Convention moved to apply for a hundred low power stations through local churches that would be the actual licensees, seeing them as the basis of an alternative network. If low power does develop (it is unclear whether it will survive as the FCC intended it), it could provide a valuable resource and alternative.

Multipoint distribution service (MDS)

MDS began as a way for apartment buildings, hotels, and other high density locations to receive special television signals for their own closed-circuit "house cable" systems. MDS has been freed by the FCC to operate any way it wishes, and MDS now offers Home Box Office and other subscription channels to individuals in cities without cable. Cost limits its usefulness to high density settings and it cannot compete well with subscription cable.

Home video

Home video is the term used for either video cassette or video disc services available in the home. A breakthrough came in 1975 with the introduction of Sony's Betamax video player, a video tape player using one-half inch video tape cassettes. It is a relatively inexpensive way for viewers to have more control over TV fare. With video recorders, viewers can play either fairly inexpensive prerecorded tapes (movies, adult films, and educational shows are the top sellers), or can record off-the-air programs for later viewing. This is the main way video recorders are used. Their prime use is to record one show while watching another on a different channel so both can be viewed or to automatically record while the users are asleep or away from home so that they do not have to miss the programs they want to see. This use of VCRs (video cassette recorders) has led to their being called "time shift devices" in the trade.[6]

The other advancement in home video has been the video disc player. This unit resembles a record player and the discs are like records. The advantages over video tape are cost (the records are cheaper than recorded tapes) and flexibility of information storage. Owners can buy such things as encyclopedias on records, flipping through the "pages" to the one they want, which appears on the screen. Disc units cannot record off the air, all the discs are prerecorded, and this prevents them from competing directly with VCRs. Their future is unclear, partly because the four competing formats of discs are not compatible with one another.

Direct broadcast satellite

Direct broadcast satellite (DBS) is a service that will carry TV signals directly into the home through use of space satellites and rooftop satellite antennas. DBS, like another network, will offer subscription cablelike programming. Subscribers will pay a fee (estimated at $100) to be connected to service for which they will receive a rooftop antenna and a converter box that will hook directly to their TV. For a monthly fee of $25, they will continue to get several channels of programming.

The Communication Satellite Corporation (Commsat) proposed in 1980 that they begin DBS service with four satellites and a "remote addressable" terminal in the home. This remote addressable feature means that the satellite can control each individual home terminal, shutting off access to those homes that have not paid and billing subscribers for each show viewed.

The FCC will have to consider how to license such a new service (Is it broadcast or cable?) before it can begin. All observers agree that DBS has a relatively short time within which to enter the market and make back its investment before the inevitable wiring of the nation to cable or other services occurs. It is clear, however, that licensing a service such as Commsat's to provide DBS service now means that it will be a programming source in the future as well. Once all homes are wired to cable and the satellite reception capability is no longer necessary, DBS programming will shift to cable and become another "pay cable" offering.

Security, banking, shopping

Security services are already one of the most popular advantages of cable television service. Subscriber homes can be wired with a number of sensing devices for burglary, fire, or accident detection, and can be hooked up through the cable system with police, fire, and private security firms. Eventually, utility meters will also be read through the cable.

Many Americans already bank by phone, paying bills with bank-written drafts commanded by digits punched into their pushbutton phones. Others also use electronic funds transfers (EFTs) that do away entirely with checks and cash. In the future, cable will allow customers to bank through their TV sets, displaying balances and other information directly on the screen.

Once banking by cable is in place, shopping by cable will also be possible with buyers executing purchases and payments from home.

Already, QUBE subscribers in Columbus can watch special "infomercials" (illegal in broadcast television), whole programs devoted to selling goods and services that can be ordered by pushing buttons on the television terminals. Payment is made by EFT debits to subscribers' charge card accounts.

Teletex, videotext, computer publishing

Cable and other transmission pathways into the home (including conventional broadcasting) can carry things other than video programming. Among the most interesting potentials in this area is the capability for information in alphanumeric form (letters and numbers) to be carried on the television screen. Controversy continues over which system is best and exactly what to call this video/computer communication service, but by the end of the 1970s two different forms of it had evolved, both of which may be available in the near future.

Teletex is the simplest form of this service and is a one-way system. In its current form, in use in England since 1976, it is called Ceefax, and involves the installation of a converter unit on the user's television set. The Ceefax central computer then stores any number of "pages" of alphanumeric information, such as lists of current movies in town or lists of important phone numbers or services. When a user wishes to have access to one of those "pages," its number is entered on the adapter on the television set, and the page appears on the screen. This system is rather slow, because it transmits pages over part of the television signal not used for video or audio and the space there is limited. The area used is what is called the "blanking interval," the dark bar that appears between the pictures on a television screen that is out of horizontal adjustment. *On Computing* magazine projected that by 1985, five million Ceefax sets would be in use in England.[7]

A more sophisticated form of this service, called videotext, is a two-way system. (The difference between the two systems is not evident from the name, so it will undoubtedly change.) After its original success with Ceefax, the British post office began experimenting with an interactive teletext system, one that would allow the user to request information, and direct the formation of it through the teletext system. The current form of videotext in Britain is called Prestel; in France it is called Antiope. (CBS is interested in introducing the French Antiope system into the United States. It is more expensive for users and for information providers—the companies and agencies that purchase space on the "pages.")

Videotext systems differ from teletext systems in one important technical way. Prestel and other advanced services are not bound by the limitations of the "vertical blanking interval" because they use telephone lines to convey their signals and thus can be directly "two-way." Users can "browse" through pages, stopping at the one they want. They can be led to a basic directory page, and then choose from it the additional pages they want to see. The system can manage educational curriculum in which users answer questions on the screen and then are channeled into different tracks of information or training.

To offer service on videotext, information providers in the Prestel system must pay an initial fee for a certain number of pages over which they have total control. They can then charge whatever additional amount for those pages they deem appropriate. The user is alerted to the cost of access to given pages and is charged by the phone company, which manages the system, for those pages. The information provider receives the money thus collected, after a small fee has been charged by the phone company. The user's cost for Prestel is about $2,000. This purchases the special Prestel adapter for the television set. When users wish to hook up to the system, they dial the local Prestel number, for which they are charged at the normal phone rate, and then pay for pages viewed at about four cents a page. They can play video games with some information providers, order goods and services from others, and obtain written information from still others.

Likely to be an early user of a videotext would be the telephone company, which could use it to provide the telephone directory without the expense of printing and distributing it. Indeed, the successful introduction of videotext in France was carried out when the phone company stopped printing the books, forcing subscribers to use videotext to search for phone numbers. Phone companies are also interested in videotext for expanded advertising in the Yellow Pages. Yellow Pages ads could be updated daily, even hourly, vastly increasing their income potential. Newspapers would be threatened by that development because much display advertising would go to the Yellow Pages, cutting into their income. The phone company is presently prohibited from entering this field, but it may soon be free to do so.

Newspapers are interested in the development of "computer publishing." Writers "write" on computer terminals (called VDTs or

CRTs for video display terminals or cathode ray tubes) and pages are stored in data banks until called up by readers. Already, documents are available in this form through suppliers such as the Compuserve Corporation of Columbus, Ohio, which offers, among other things, 11 regional newspapers, Associated Press news and sports, and a variety of information services through computer-to-home-computer-linkups.

The *Columbus Dispatch* was the first newspaper to be distributed by computer. Subscribers who wish to read it in that form do so by calling a special phone number, hooking up the phone to their home computer, and then scanning the stories in the paper on their VDTs. If they wish to save a story for later, or have a copy for their files, they can order a copy printed by their home-computer printer. Compuserve and other such networks also offer expanded data operations for home computer owners through phone hookups with their central computers, and also offer a kind of electronic mail service, which allows two subscribers to leave messages for one another in the computer that will be transmitted to their home terminals electronically.

The newspaper and magazine industries are looking at the potential of computer publishing. When they move into it, it will be relatively easy for them. Most major newspapers are already writing, editing, and filing stories by means of VDTs, and it will be a relatively minor shift for them to distribute stories by computer, eventually eliminating expensive printing, distribution, and mailing costs involved with the conventional newspaper.

Even the U.S. Postal Service is interested in forms of computer publishing and videotext services. First, any developments that would allow the phone company, for instance, to enter that field, would be a tremendous threat to conventional postal services. Second, the Postal Service itself is interested in being the provider of such things. Already the Postal Service is studying "electronic mail" which would allow businesses to send letters electronically. The letters would be received by the post office in the city of destination, printed there, and delivered with the next day's mail. Eventually, the Postal Service would like to eliminate hand delivery altogether, moving to a full electronic link that it would provide. Users of Western Union's "Mailgram" service already have an idea of what that would be like. With electronic mail, the letter would be generated (typed) by the sender, and would be transmitted over phone or cable links to its destination where it could either be displayed on or printed out for circulation.

The Postal Service may be prevented from entering this field because of what private industry sees as unfair competition from government. The Constitution does direct that the post office be involved in mail delivery, however, so a long debate looms over whether "electronic mail" can be called "mail" in the Constitutional sense. Eventually, hand delivery of mail on a daily basis will be replaced by daily (even Sunday) delivery of electronic first-class mail; handcarried delivery of packages and other classes of mail will be reduced. Magazines and newsletters wishing to avoid such delays in delivery will "computer publish," becoming instantly available on an almost constant basis to subscribers over their home video screens.

High resolution television and improved audio services

American television suffers from one major technical limitation. Because television was developed and marketed early here, the inventors and producers settled on a standard for technical transmission (with the involvement of the FCC) that is inferior to technical standards elsewhere. U.S. television video image conforms to a standard called NTSC and contains 525 lines of resolution. (Lines of resolution are the hundreds of little horizontal lines that make up the picture and are visible very close to the screen.) These lines are the glowing paths of an electron beam shot from the electron gun at the back of the picture tube, constantly writing new pictures on the screen.

In Europe and other places where television emerged later, other, more sophisticated standards were developed. Those standards were of either 625 or 725 lines of resolution. Many more lines are thus packed into the picture, increasing its quality and definition. One reason that projection television systems that blow up the image to several feet diagonally are not more popular (besides their cost) is that when a fuzzy image is blown up, it becomes a fuzzier image.

Scientists and researchers have been working on this problem for years. In 1980, RCA announced a breakthrough, a new kind of screen that could be used with regular 525-line transmission signals and would deliver a high resolution, 50-inch diagonal picture through a screen one inch thick (instead of that long picture tube) and hung on a wall. RCA projected that this screen would be available by 1990.

Another area of research in resolution is what is called "high resolution television," a technology that will use 1,000 or more lines of resolution, thus improving quality. Such equipment will be com-pletely incompatible with old 525-line systems and will require an en-tirely new network of cameras, transmitters, receivers, and sets. One of the possible uses for direct broadcast satellite services would be for them to provide such high resolution signals, as they will be able to bypass all the conventional 525-line systems. Such a service will be very expensive.

DBS, disc, and other services will be used to improve the quality of audio into the home, both for television and for hi-fi and radio listening. One plan for both DBS and video discs is to offer 18-track recordings in flat volume, which listeners may adjust as they wish for the right balance of each instrument on the recording.

Home information centers

The new age will be marked by the convergence of technologies. In the average home, this will culminate in the in-troduction of the home information center. This device will incor-porate telephone, video, computer, security, mail, news, and in-teractive services in one unit that will be connected by cable or fiber optic phone lines with larger computers elsewhere.

There are no pictures of what these centers will look like yet but they probably will not be much different from equipment available already. They will be nearly identical to computer terminals, telephones, and televisions that are already familiar. What will make them unique is that they will represent, for the home, the con-vergence of technologies. All the services above will be available directly to the home through high capacity transmission. Whether or not the home becomes one giant "Buck Rogers" control center, the home will become one component of a larger network of com-munications, data, and commercial activity in a way it never has been before.

Often mentioned as one outgrowth of this "center" will be the ability of many people to work from home using a home computer or an employer-supplied terminal. Some of the advantages to employers are obvious. A large work force can be employed in infor-mation services with little overhead. Management of activities can be handled through the computer, and the supervisors can have direct contact with employees by traveling to the employees, rather than vice versa. Many people already work from home in professions such as survey taking, telephone sales, and claims adjusting. The sophistication of the new technologies will allow an.expansion of the range of activities possible from remote terminals. It is not likely that everyone will work from home even far in the future. It is likely, though, that more people in more classes of occupations will do so as homes come "on line" with broadband (high capacity) interconnection to the North American communication/data grid.

The earlier age might well have been called the age of broadcasting, marked by the prominence in society of broadcast services. So this new age can be called the age of broadband with homes linked to the information network.

But what will the effects of these developments be on the homes hooked up? What major issues will surface as the age dawns?

Already, the introduction of new services in homes in areas where they are available means spending more time with television. This is true with home video, and research indicates that it holds true with cable subscribers also. People who have more channels to watch tend to watch more. Conventional broadcasting does not seem to suffer much; it is watched anyway. Cable subscription channels are watched in addition to normal viewing. If one concern in the broadcasting era is the amount of time people spend with television at the expense of family and interpersonal contacts, then the future is more alarming. People will watch and interact with their hardware more, and with other people less. At least that is the way trends look now.

Dr. Marc Porat has suggested three other areas of concern as the new age comes.[8] One was predicted in John Wicklein's 1978 study of the QUBE system published in *Atlantic Monthly*. As discussed earlier, Wicklein found that privacy was the major emerging problem with the new technologies. In even so rudimentary a system as QUBE, subscribers can find themselves the subjects of sophisticated files on their viewing, buying, and financial habits. With the future home information center, centralized data banks will have collected, as a matter of course, nearly all vital information about each subscriber.

Wicklein found disturbingly little thought being given to the issue of privacy. The problem is not that some sinister force necessarily wants to gather these data. Instead, the new age will bring the inadvertent collection in one place of data that, while available before, had never before been centralized. A computer system that manages a household's phone calls, mail, entertainment and educational viewing, banking, purchases, and home security knows more about that home than any one source has ever known before.

Marc Porat contends that this is a tradeoff that may have to be made:

> It is surely the case that we may have to surrender some of
> what we have called privacy in order to take advantage of new op-
> portunities.[9]

Porat suggests two other areas of concern about the convergence of technology, the problem of "information overload" and
the problem of the "information poor."

Overload suggests that, when people have access to all these
technologies, they may simply have more information at their
disposal than they are capable of handling. How does a person
manage a cable service with 104 channels? A channel or two will be
needed just to list what is on the other channels. People will have access to so much information that they may be worse off than they
were without it. Unless, that is, they develop new methods for cop-
ing with and using this abundance.

The problem with the information poor is stickier. Already, increasing numbers of Americans (although not an increasing percentage) are functionally illiterate. How will they move into an age that
presupposes basic skills for access to basic services and inputs?

The financially poor are also candidates to be information poor
because of the threshold costs of these technologies. Certainly nearly
everyone has television. Fewer, however, will be able to afford cable,
subscription cable, or home video. Still fewer will be able to have
home computers. Unless provisions are made, the mere development of the new age could vastly expand the size of the U.S.
"underclass," now defined not only by economic status but also by
access to society's mainstream of information.

The convergence of technologies now occurring will harm some
traditional relationships in other ways. In spite of its ills, conventional
broadcasting, through its public service mandate, has provided information, education, and support to some who would not have had it
otherwise. The logic of the new age is "pay for play." As older services are supplanted the risk is real of losing some things of great
value.

Public broadcasting undoubtedly will also suffer loss of audience
support and programming. (That started with the BBC decision to go
to pay cable in 1983.) The people who can afford to subscribe to
public TV and volunteer time for its fundraising activities can also afford to subscribe to pay cable where they can get the programs they

want for themselves and their children with less hassle. Public TV will be poorer for that.

Libraries and other local institutions will also be changed by these developments. Many libraries certainly will offer their services directly to the home via cable, enhancing their role as an information supplier. The library's role of maintaining culture through the promotion of literature will be diminished, however. Books may not go out of fashion all together, but they will be used less, making the libraries' "information service" role potentially their only role.

One of the few institutions that has already benefited from the new age has been religious broadcasting. The technological developments in the 1970s placed radio and television religion in an especially advantageous position. The next chapter will look at this phenomenon as a case study of one of the early manifestations of the new age of telecommunications.

One of the dangers of this transformation into the new age is that it will happen without most people even knowing about it. These developments are profound but subtle. Their very success depends on their being introduced with a minimum of disruption. One way of seeing the extent to which this new age has already arrived is to take a personal "telematic inventory" that catalogs the aspects of the new age that are already available. In the following list of questions, score the number of points listed for each yes answer.

1. Do you have a pushbutton phone? (1)
2. Can you bank by phone if you want to? (2)
3. Do you have a Visa or Mastercard account? (1)
4. Do you have a color television set or more than one television set? (1)
5. Do you have direct deposit of your paycheck to your bank account? (2)
6. Are you able to purchase travelers' checks or obtain cash from an automatic teller machine? (1)
7. Can you dial a joke, the weather, sports scores, or some other service for a small charge to your phone bill? (1)
8. Do you subscribe to MCI or any other non-Bell long distance telephone service? (2)
9. Do you have a pocket calculator? (1)
10. Do you have a pocket computer or programmable calculator? (2)
11. Do you subscribe to cable TV? (2)

12. If yes, do you subscribe to "pay channels"? (Score 2 points for each channel.) (2)
13. Do you have extra services available with your phone such as call-waiting or call-forwarding? (2)
14. Do you subscribe to multipoint distribution service (HBO or another cable channel but no basic cable service)? (1)

Scores
1-5 —telematic Luddite
6-10—a bit backward
11-18—'on the way'
19-up—a real progressive

These scores are not intended to be unduly stereotypic, but rather to help individuals assess the degree of their own participation in the telecommunication age. All these services require packet-switched computers. The financial services require limited EFT-like functions, and the hardware mentioned are significant of the kinds of hardware that will typify the new age.

Chapter 10

The Electronic Church: Vanguards of the New Age

In the early 1970s, a young man working in a small-market radio station as the Sunday morning shift announcer (the bottom rung) first became intimately aware of the phenomenon of broadcast religion. Although he remembered having seen Oral Roberts and Billy Graham on television as a child, he had tended to disregard their power or effect as communicators. He had thought Roberts in particular seemed almost humorous with his tent-meeting, faith-healing atmosphere. Many people in the young radio announcer's own church *were* impressed by Roberts, however, much to the consternation of their pastor.

But the radio preachers he heard in the early '70s were different. They spoke to their audiences as though to a regular congregation, offering to communicate by mail and to answer prayer requests. New television preachers began to appear, such as Jerry Falwell from Lynchburg, Virginia, and a "new" Rex Humbard. Even Oral Roberts reappeared in 1973, now a "mainline" minister, and with a radically changed format.

Roberts' transformation was perhaps the most typical of the changed face of religious broadcasting in the 1970s. Gone were the old tent and folding chairs. In their place were a studio and neatly dressed audience. Roberts himself was nattily dressed and surrounded by fresh young faces from his college, singing his theme, "Something *good* is going to happen to you." Cynics declared that he had made a bid for respectability through his university and new TV image. But whatever he did, he became a success and is still one of the most successful of television preachers.

By 1973, Oral Roberts, Rex Humbard, and Jerry Falwell had discovered the power of television for creating an audience (a congregation) through regular contact and through production values

that stressed the personality of the preacher rather than the experience of the service. Watching these programs and listening to the
radio preachers provided the sense of a subculture or community
that gathered around these ministers every Sunday. The cacophony
of ads and pop music and news that dominated the schedule at the
station was replaced for a few hours by a cacophony of preaching,
counseling, a different sort of selling, and a different kind of "pop"
music.

The music was an important clue to developments that would
bring electronic religion into prominence by the end of the decade.
There were hundreds of recording artists. A complete subculture of
writers, producers, performers, promoters, and packagers emerged
that in many ways rivaled Hollywood in sophistication. There were
jazzy instrumental renditions of "Just A Closer Walk With Thee" by
Billy Graham associate Ralph Carmichael. There were rhythm-and-
blues renditions of the same tune by Andrea Crouch, a performer
who gained critical acclaim outside the religious realm as well. There
were breathy, sultry, almost sexy renditions of "He Touched Me" by
Anita Bryant. Gone were the days of the tacky, glad-handing, used-
bus-tour singing family, hawking their own vanity recordings. Even
the Happy Goodmans appeared on a "major" label. Religious music
had hit the big time, and hit it solidly when the main label, Word
Records, was sold to ABC in 1976.

That religion had become a marketable commodity was confirmed when the Gallup organization began publishing a "profile of
the Christian marketplace" late in the decade. This trend accom-
panied a general trend that gained secular attention in the mid-1970s
when evangelical Christianity emerged as a major religious and social
force. George Gallup suggested that 1976, the year when an
evangelical was elected president, should be called "the year of the
evangelical," according to *Newsweek*. The same article called
evangelicalism the most significant and overlooked religious
phenomenon of the 1970s.[1] A more studied observer, Dean Kelly,
analyzed the effects of these trends on churches in the book
Why Conservative Churches Are Growing. Evangelical churches, he
found, were growing in numbers, growing in middle and upper-class
membership, and growing in self-identity.

These newly aware evangelicals were the market for the records
and radio programs, but they also became part of a new development in television service as well. In 1974, an entirely new television
station went on the air in San Francisco. This was to be a *religious*

television station, devoted to full-time religious programming. At the time, this seemed an impossible task because a limited number of programs was available.

The new station manager confessed that he was desperate for programming. Most shows he could use were available on videotape or film and the cost of shipping them by mail was going to be a problem. What time they could not fill with these "bicycled" shows (films and tapes shipped by mail), they planned to fill with their own locally produced shows. These shows were a mixture of traditional formats, including talk shows and a children's show with an ersatz Indian chief who urged kids to write in for "free things." The manager expected that many of his problems with filling time would be solved when new program suppliers in the East obtained access to satellite channels so that live, high quality programming from elsewhere could be added to the schedule relatively cheaply.

By the next year, a person living in the East could see the kind of programming to which the manager had referred. Pat Robertson's "700 Club" was then available in major cities, including Chicago. When such shows were a fixture only in the "Bible belt" and rural areas, they made relatively little news. But when they invaded the "theological education center" of the country, Chicago, they achieved some attention in the theological community. By 1979, even so prominent a commentator on the Protestant theological scene as Martin Marty of the University of Chicago was moved to remark in writing about their potential impact:

> The electric church isn't a denomination—yet! But it threatens to replace the living congregation with a far-flung clientele of devotees to this or that evangelist or entertainer.
>
> Composed of customers on the receiving end of religious television and radio empires, the electric church is supported by people who are attracted to a religious leader, send in weekly or monthly checks and pray for the cause. . . .
>
> This invisible religion is—or ought to be—the most feared contemporary rival to church religion. . . . [2]

It is not clear that such a dire prediction as Marty's is justified nor those of more persistent critics such as Dr. William Fore of the National Council of Churches, who calls the phenomenon the "electronic church" and defines it this way:

> By the term electronic church I don't mean *all* broadcast-

ing done in the name of religion. I mean specifically those programs that present a preacher and a religious service and that are
aimed at creating a strong, loyal group of followers to that
preacher and service.[3]

The jury is still out on the case of the electronic church. What is
important about this development is that it is an important first
development of the telecommunications age. The electronic church
is a phenomenon that exists precisely because of the forces of
technological change. Without satellites, cable, and computers, the
electronic church as it is known would not have developed. Tracing
its history permits tracing the emergence of forces that will fun-
damentally change the information society in other areas.

Religious broadcasting is nothing new. Charles Swann reports,
in his study of the electronic church, that the first voice broadcast in
history was a religious service sent to ships at sea in 1909. The first
radio church service was broadcast in December 1920 over KDKA in
Pittsburgh one month after the initiation of commercial broadcasting.

In the early years, both evangelicals and nonevangelicals saw
radio as a powerful force and moved to use it. Many evangelicals
received licenses for radio stations in the 1920s as did some
nonevangelicals.[4]

The initiative, however, moved to the nonevangelical, mainline
churches through the period of the establishment of the FCC between 1927 and 1934. In exchange for promises made to Congress
to provide churches and other public groups with free air time,
broadcasters achieved legislation that did not bind them to such activity or force a portion of their broadcast day to be turned over to
public groups, which had been proposed. The well-organized national groups, which tended to be mainline Protestant and Catholic
(later joined by Jewish groups), sought to hold the broadcasters to
their word.

The system that developed was a relationship between the major stations, and later the networks, that guaranteed sustaining (free)
time to these national ecumenical, dominant groups. Nonmainline
groups, primarily evangelicals and independent broadcasters, were
left out of this sustaining time system. In his history, Ben Armstrong
reports that evangelical groups, at that time on the religious "fringe"
compared to the mainline agencies such as the Federal Council of
Churches (later the National Council of Churches), were left out of
the sustaining time system. Because the system was encouraged by

the FCC, pressure increased to eliminate early paid-time religious broadcasters, including some who bought network time. As this occurred the evangelicals became increasingly alienated from dominant religion, lost some station licenses they had held, and reverted almost entirely to locally syndicated, paid-time broadcasting.

This dualistic system held well into the 1970s: mainline ecumenical Protestants, Roman Catholics, Jews, and the Southern Baptists dominating network and local large market "free time"; independent evangelicals dominating "paid time." In 1978, the National Council reversed its policy on sustaining time, for the first time giving approval to member denominations to purchase time.[5]

By then the quality and quantity of sustaining time available to the ecumenical groups had deteriorated because of commercial pressure and pressure from the independents; many groups wanted to buy time to compete.

Sociologist Jeffrey Hadden has suggested that the paid-time tradition of the evangelical independents put them in position to capitalize on expensive communication technologies and new developments.[6] They had become used to the idea of having to spend money and were able to openly raise money on the air. The mainlines on the other hand, had become used to getting access on sustaining time, and for reasons of propriety, theology, and agreement with their host networks and stations, did not openly raise money through on-air solicitation. When technology caught up with the system, the evangelicals were in a position to move; the mainline ecumenical groups were not. While these institutions had a tradition of funding publications and church school curriculum and selling them, similar ways of doing broadcasting did not develop.[7]

Hadden thinks certain technological developments are keys to understanding the emergence of the evangelical broadcasters because of the broadcasters' willingness to pay for these technologies. Primary among the technologies he sees involved are computer-based direct mail capabilities and new distribution networks, primarily satellites. The earlier development of the UHF television service could be added to the list. Until fairly recently, UHF was largely unprofitable. It provided new outlets for religious broadcasters at low cost; they purchased either air time or whole stations. The other important development has been the increasing number of cable television systems with empty channel capacity. Many of these can receive satellite signals and thus can easily add a satellite-fed "religious" channel.

One further historical factor has been the role of the Federal Communications Commission regarding these broadcasters. Although the FCC inadvertently encouraged the independents to purchase time, and the direct Commission involvement with religious stations has been important in such cases as the Supreme Court's Red Lion decision, Commission activity has been uneven. In a 1978 article in a prominent law journal, Linda Jo Lacey suggests that Commission policy on religious broadcasting actually encourages independent, evangelical broadcasting by refusing to contemplate it at all.[8] Commission reluctance has been encouraged recently by the phenomenal case of the Lansman-Milam petition which, in spite of the fact that it was refused by the Commission in 1975, continued to stimulate thousands of protest letters a year well into 1980.

The Lansman-Milam petition, as described earlier, was not a petition with signatures, but a legal document requesting action on a rulemaking by the FCC to look at whether sectarian groups should be allowed to hold FM noncommercial licenses. The petition, RM 2493 (RM for rule making) was prompted by honest intentions. Jeremy Lansman and Lorenzo Milam were consultants who worked with minority and nonprofit groups to obtain noncommercial radio licenses for broad community uses. They had become concerned that sectarian religious broadcasters, who could afford to go for regular licenses because they raised money on the air, were snapping up the cheaper noncommercial licenses. Lansman and Milam contended that such sectarian uses of these licenses propagated narrow, conservative voices and did not cover well the broad expanse of religion.

The sides became antagonistic and the evangelical trade organization, the National Religious Broadcasters, began raising a war chest to fight the petition at the FCC. Some NRB statements indirectly linked the Lansman-Milam petition to atheist Madalyn Murray O'Hair, who denied any involvement, and things were off and running. Such a petition, seen as a "petition against God" promoted by the country's leading atheist, was bound to raise hackles and money in the hinterlands. The petition drives spawned by this incident continued even after the FCC turned back the petition in 1975 without comment, and seem impossible to stop. Whenever the furor seems to have died down, someone picks up an old letter or petition and, thinking it is new, starts things again. The FCC has set aside a room in its building to hold the letters and has stopped writing back.[9]

How do the electronic church broadcasters use the new technologies to their advantage? Ben Armstrong, executive of the NRB, reports in his book, *The Electric Church,* that the weekly audience for religious radio and television is over 130 million. Though this is most certainly an aggregate figure (counting each person over again each time a different program is tuned in) and overestimates the number of people who regularly watch and listen, the audience is large. A study done in 1972 found that in an average week, 47 percent of the American population turned on at least one religious broadcast. The *Wall Street Journal* of May 19, 1978, reported that at that time one new religious radio station was going on the air each week and one new television station each month. These trends are only the latest developments in a field that is still dominated by syndicated producers (ministries distributed nationwide rather than only in a local area), at least three of whom now distribute their programming by satellite, according to Armstrong.[10]

Probably the most controversial issue is the funds these broadcasters raise. Charles Swann, in his study of religious broadcasting, estimated that the yearly income to these ministries ranges anywhere from $12 million to $16 million (for Robert Schuller's Garden Grove Community Church in California) to over $60 million for Oral Roberts.[11]

Although most critics and observers of the electronic church assume it to be a monolithic movement, this is not the case. A brief rundown of the most prominent ministers reveals many differences. Several are based in local churches. (The most prominent are Jerry Falwell's Thomas Road Baptist Church and Robert Schuller's Garden Grove Community Church.) One of these, the Worldwide Church of God, began as a radio program with Herbert W. Armstrong and became a denomination, represented now on television by his son Garner Ted Armstrong.

All of the electronic ministries are first and foremost syndicators of programming, usually on a paid-time basis. One, the Christian Broadcasting Network (CBN), is also a station owner, holding four UHF television licenses and five FM radio licenses. All of them syndicate either by "bicycling" or by direct surface feeds. CBN and the PTL network lease 24-hour-a-day satellite channels, PTL feeding its own programming exclusively, CBN subletting capacity to others.

Most of the electronic ministries rely on direct mail for support and thus have the latest sophisticated direct mail computers. Some also have toll-free numbers that people can call for "counseling" referrals and for pledging donations. CBN and PTL also have regionally located call-in centers for followup. CBN, according to its founder, Pat Robertson, staffs 71 such centers with professional directors and volunteer staff.[12]

Another feature of many electronic church ministries is publishing pamphlets, books, tracts, and monthly magazines. Many of them regularly mail material to their "faith partners," sending additional materials or religious articles to those who give more money. Oral Roberts sends models of his university's prayer tower to contributors. Robert Schuller offers a "gift of the week" to anyone who will write in (to be placed on the mailing list). While Schuller was building his new church, the Crystal Cathedral, he offered small models of its windows to contributors of over $500 whose names would be placed on plaques in the actual windows of the church in the manner of European cathedrals.[13] The PTL Club calls its more generous donors "faith partners" and, according to a published article, offers different levels of "partnership," signified by different gifts the pledge will bring:

> . . . a PTL stickpin, the PTL newsletter, a PTL Bible, and for a one-time gift of $1,000, an heirloom Bible, illustrated, with a stamped antique gold cover with the donor's name printed on it.[14]

Managing these nonbroadcast contacts is one of the primary accomplishments of these ministries and one of the ways in which they are indicative of the new age of telecommunications. Hadden details how this works:

> Phone banks with toll-free numbers and/or regional centers can process hundreds of calls during a telecast. And each phone call, whether the caller is making a financial contribution, requesting gratis materials, or seeking counseling and prayers, is another name about to be readied for a variety of appeals and solicitations. In a short time leaches can be filtered off magnetic tapes and repeat contributors identified and acknowledged with seemingly personalized attention. On-line printers mass-produce letters with the receiver's name printed throughout the text as if it were written and typed individually for that person. And key lines can be underscored seemingly with the pen of the signature. For modest contributions one can designate a "faith partner" or some other title. For a little more, one is offered a membership in "elite" clubs, and potentially big spenders can be identified and singled out for genuine personalized attention

like retreats, weekend seminars, and the like.

People in the electronic church encourage their listeners and viewers to write or call and share their personal problems and needs. The more personal information available, the more responses can be targeted to individual needs and interests. If someone in the caller's family has a drug problem, then there are tracts and letters about drug abuse and what to do about it that will soon be in his or her mailbox. Marital problems, difficulties with children, fear of crime in the street, difficulty in managing money, whatever the individual's problem, chances are that enough other people have the same type of problems that materials have been prepared.

The mail rooms of the more successful members of the electronic church are paragons of modern communication technology. Mail is sorted by the presence or absence of money. Then, letters are sorted out by topics and appropriate paragraphs are retrieved by computer and woven into an appropriate and "personalized" response. Then, on-line printers dash off these individualized responses. Oral Roberts, who frequently tells his audience that he answers every letter, reportedly has a mail room capability for processing 20,000 letters a day.[15]

The availability of satellite channels, which solve the programming problems of stations like the one in San Francisco, and the sophisticated "back room" techniques Hadden describes, are only part of the story. Access to UHF television stations that are operated as Christian outlets and access to spare cable channels have made the electronic church the phenomenon it has become.

In a real sense these ministries (and there are hundreds besides the prominent ones that have been mentioned) have pioneered the way for other alternative uses of television. They have shown that it is possible to sidestep the traditional networks to gain a foothold in the media world.

The electronic church is not without critics, and many criticisms are justified. To attract and hold an audience, these broadcasters have mimicked the methods of television and have made their preachers "stars." Serious questions may be raised about their ecclesiology. Can they claim to be the "church" or, in Christianity, is that term reserved for physical gatherings of the worshiping community? There is much misunderstanding on both sides of the issue.

Many of the broadcasters and their critics came together for a consultation in February 1980. As an outgrowth of that meeting, research may be done to answer the questions to which no one yet has answers—who watches, why, to what effect.

What is clear is that in the last years of the decade in which they emerged, some of these ministries became political forces as well. In April 1976, *Sojourners,* a politically liberal evangelical magazine, reported on a movement called "the plan to save America" which involved Pat Robertson and a number of prominent evangelicals working with political conservatives on a coalition to move America to the right. The *Washington Post* report on the 1980 NRB convention (January 22, 1980) noted that President Carter spoke and, among other things, reported asking China's premier, Deng Xio Ping, to once again open China to Christian missionaries. The *Post* also reported that other major presidential candidates visited the convention, and over 100 members of Congress were involved in the NRB's "Congressional breakfasts" during the week.

During the 1980 election campaigns, Jerry Falwell emerged as a political figure, founding an organization called the Moral Majority. His political views had been well known. In an article on Falwell for *Esquire* magazine in 1978, Monica Murphy reported that his Liberty Baptist Bible College was strictly run as an alternative to the anti-American, secularizing trends he saw in most educational institutions. She quoted him:

> Developing Christian character is our main purpose. We take our students from kindergarten through college, so we can shape them. We find that when they grow up that way they don't revolt.[16]

At the time of that interview, Falwell had already begun a campaign to "clean up America." He had made up 50 million ballots asking people's opinions on abortion, homosexuality, and pornography. They were distributed through paid advertisements in national magazines; reportedly he expected 90 percent of Americans to be against those things. He does not wish to be a political leader, according to Murphy, saying, "If I'm too political I'll lose my effectiveness." Asked what the differences are between political and moral issues, he responded, "The Panama Canal is political, the ERA is moral." He does not eschew politics entirely, having campaigned with Anita Bryant in Dade County, attacked President Carter for giving an interview to *Playboy* magazine, and campaigned in California on behalf of an anti-gay rights initiative. Bryant stood beside him in a 1979 program when he called for a return to ". . . the McCarthy era, where we registered all Communists." He went further, suggesting that they not only be registered but that " . . . we should stamp it on their foreheads, and send them back to Russia."[17]

Falwell's 1980 activities were limited to his involvement with the Moral Majority, but many observers were waiting to see how prominent a role he would play in the Reagan administration elected that year.

Falwell's political influence is a creation of the technological times. His prominence is related to his ability to preach to millions of homes over television. It is important to remember that television's effect is not so much its potential for people's being influenced or brainwashed by persons such as electronic church preachers, but its ability to coalesce an audience around such figures. The audience may be drawn to the preachers for any number of reasons, but once there it can be organized around certain principles and activities. The audience may not be made to agree with something, but once there it becomes a foundation on which political action can be taken. It is an axiom of the new age that the flexibility of the new media allows those who have the money to buy access to audiences that never existed before.

What is the future of the electronic church? Many more programs are competing for air time almost daily as the new age gets underway. Partly as a result of the "success" of the electronics, some mainline churches are now stimulated to go on the air but in slightly different ways.

The Church of Jesus Christ of Latter Day Saints (Mormons), already a major owner of commercial broadcasting through Bonneville, Inc., has begun installing a satellite teleconference system that will link all of its temples with the parent church in Salt Lake City.

The United States Catholic Conference plans to lease satellite capacity to link Catholic schools, dioceses, and offices for distribution of educational and program material. The Catholic Communications Collection, a special offering for communications begun in 1980, should give Catholics the capital needed to move into new technologies. The National Council of Churches has initiated studies of satellite and cable possibilities, and the largest U.S. Protestant denomination, the Southern Baptist Convention, has announced plans to construct a satellite distribution system. The second largest denomination, the United Methodist Church, has announced plans to actually purchase a broadcast station, the basis of entry into new technological distribution.

What will these developments bring? George Conklin, a prominent researcher in religious communications, has this assessment.:

> These institutions will be changed by their use of these technologies. They will be changed in ways so profound yet so subtle they may never know how they are different because of them.[18]

Ten years ago, seers might have projected that the church would be the last institution touched by a new technological development. That it has been one of the first is indeed surprising. The major effect on the mainline or dominant church has been a feeling of being outflanked, threatened. Some profound mistakes could be made if these traditions do not look critically and creatively at the opportunities these technologies offer and, instead, rush to compete.

One of the most profound differences between mainline church broadcasting and the electronic church is in the way the audience is seen. Mainline churches see their audience and their constituents as different groups, addressed differently. Constituents are asked to support programming that serves general religious needs of the country and that puts a religious voice on the otherwise secular air.

Electronic church ministries confuse the distinctions between constituent and audience so that they become the same. Audience members are encouraged to support the extension of electronic church ministries; they become constituents. The electronic church is not accountable to the contributor, whose only accountability mechanism is the process of viewing and contributing. As long as this situation exists, mainline churches will not be able to "copy" the electronics (whether or not they really want to). The mainline churches need to use a studied, pragmatic approach, evaluating the real advantages available in the new media so they can act accordingly. They could probably not "beat the electronic church at its own game"; they probably would not really want to.

Section IV

Agenda for Reform

Introduction

Technological change clearly is overtaking American telecommunications. What is the potential for changing the direction of this process?

Since 1968, a movement for constructive change in American telecommunications has been growing. This media reform movement traces its roots to the civil rights struggles of the 1960s. Out of that experience, many people came to believe that the institutions of the media, primarily broadcasting institutions, should be encouraged to serve the needs of the poor, the disadvantaged, the voiceless.

The power of broadcasting to set the public agenda and interpret information makes it an important actor in the resolution of social and cultural ills. In addition, broadcasting was given a mandate by the Communications Act of 1934 to take just that role, to become a local "public trustee."

As broadcasting developed, it became instead a "business." Largely because of the inaction of the Federal Communications Commission, broadcasting came to see itself *only* as a business. Important court decisions in the 1960s attempted to reassert broadcasting's responsibilities as a public trustee over its rights as a business to do as it wished.

The most important of these decisions was the Federal court's decision in the WLBT case that a broadcaster with an abominable record of service to the needs of the local community could actually lose his license. This was the first time this had happened, and it radically changed the picture for broadcaster and community alike. The most important feature of this case was the court's decision that the public had a right to participate in FCC proceedings. Until that time, anyone *but* the public could be a "party in interest" to a station's license application. Dr. Everett Parker, who was instrumental in that case, said, "Even the local car dealer could get in to the FCC but public, nonprofit, and church groups couldn't." The court's decision opened the formal process of opposing a license application to members of the audience of the station so that they, too, could file a "petition to deny."

The other important decision of the 1960s was the Red Lion decision. A radio station in Red Lion, Pennsylvania, consistently refused to air material with which the owner disagreed, a flagrant violation of the Communciations Act, which requires broadcasters to provide open and diverse discussion of important public issues. The owner argued that he had a right to not cover what he wished and say what he wished—a right guaranteed by the First Amendment's freedom of speech clause.

The court's ruling is instructive to anyone who wonders about the implications of freedom of speech for broadcast regulation. The court held that because broadcasting operates on a scarce resource that limits the number of stations on the air, broadcasters have a special mandate to cover as much of the landscape as possible.

In a sense, the court's argument was this. Freedom of expression, which is intended to give the public access to "free and robust debate" in a "free market place of ideas," can be limited in two ways. One way is to strictly limit the range of ideas that can be exhibited. This abridgement of expression is the one most people think of when they consider the First Amendment. Equally important, however, is the second way this freedom can be abridged; by allowing much freedom of expression, but strictly limiting the access the public has to it. To use the example of the public square: The market place of ideas is abridged when only one carefully controlled speaker is allowed in the square; it is also abridged when many speakers are allowed with no control over what is said, but the public is not allowed to go to the square to listen.

In the Red Lion case, the court decided that because of its scarcity, broadcasting ran the risk of restricting the public's access to the market place of ideas unless it actively sought out as broad a spectrum of thought as possible and allowed free access to it. It held that the right of the public is paramount in these matters. When First Amendment principles are involved, these conditions of freedom of expression must be satisfied.

Chapter 11

The Media Reform Movement at the Crossroads

As an immediate outcome of the U.S. District Court's 1968 WLBT decision, a number of groups formed across the country with the express purpose of advocating improved service of local needs and interests. Among the first areas addressed by these groups were the quality of children's programming and service to minority groups, including equal employment opportunities.

One of the first of these advocacy groups, and still one of the best known, is Action for Children's Television in Boston. ACT took an early interest in the quality of children's programming. In recent years, it has turned its attention to the amount and type of television advertising to children, primarily on "kid vid" (Saturday morning and after school programming). A significant breakthrough in quality came for many children in the late 1960s with the production of "Sesame Street," a public television program intended to give disadvantaged children a head start in school through direct instruction in basic educational skills. The work of ACT and other groups continued to concentrate on commercial television, however, as this was where most children continued to do most of their watching.

The most powerful strategies initially seemed to center around legal and legislative action. The cornerstone of this action was the petition to deny renewal, the legal document that can be filed at license renewal time in order to have input into the FCC's deliberations over a station's license renewal. Most groups found that as the three-year license renewal term drew to a close, they could begin negotiations with stations about significant issues regarding their community service, and work out ways to address those problems. Only after such negotiations failed did petitions to deny prove warranted. Many national organizations based their reform work on consulting with local groups on this negotiation-at-renewal process. The leader among these was the United Church of Christ's Office of Communication, which had brought the original petition against WLBT. The UCC was joined in advocacy work by such organizations as the National Black Media Coalition, the National Organization for Women, and the Mexican American Legal Defense and Education Fund.

Broadcasters were quite concerned about this process. Before WLBT they had had a fairly easy time with the FCC at renewal time. Even with the prospect of a petition to deny, their chances of losing their licenses were small, but the suspense and bother involved seemed an unwelcome burden to them. They argued that the petitions to deny were being used as harassment techniques by local groups and were resulting in burdensome paperwork for stations.

The petition to deny was, however, the one true sanction the local community had to obtain better broadcast service. The Communications Act's original intention—that broadcast licenses be only necessarily for three-year periods—had come to be not the rule but the exception. In practice, licenses had come to be licenses in perpetuity unless the licensee chose to sell, which was usually carried out with minimum interference from the FCC or the local community.

The WLBT case changed the rules of the game a bit, and many broadcasters appealed to the FCC and Congress for relief from this new wave of activism.

Cynics among them were joined by cynics in Congress, the FCC, and the public interest groups in observing that not much changed because of this activism. The cynics were not accurate. Many local groups found that their interest in broadcasting regulation, fueled by the fact that the petition to deny gave them redress and true standing with the broadcaster, led to real changes in local service and hiring practices. In San Francisco, for instance, the local Committee for Children's Television had great success in improving local service to children after it filed petitions to deny on that issue with local stations. In a market that had had no locally produced children's shows, two of the stations began to produce and present high-quality programming for children. This trend occurred elsewhere, and surely it is at least partly responsible for the recent increase in the number and quality of network children's shows, particularly after school. Saturday morning continued to be a problem well into the 1970s with ACT and other groups pressing further their interest in limiting the type and frequency of advertising to children.

ACT took its case to the Federal Trade Commission after the FCC did not prove helpful. President Carter's FTC, under Chairman Michael Pertschuk, began to investigate the problem of advertising to children. For a time in the late 1970s it seemed that the U.S., the only Western nation with almost no curbs on the quantity or type of advertising to children, might change that practice. The FTC staff recommended in 1978 that several limits be put on advertising: No ads directed to children under six; no ads for highly sugared products; limits on other practices in ads to children under 12. These entirely reasonable curbs (which must have sounded very attractive to most parents, who suffer the consequences of these practices at the supermarket) were immediately attacked by the industry groups involved and by Congress, which in 1979 cut off funding for the FTC to continue studying the issue. With the election of a much more pro-business and antiregulation Congress in 1980, it became unlikely that such regulations would ever be passed.

The reform movement's action on other issues also stalled later in the 1970s after a period of great activism and apparent progress. The Federal Communications Commission had taken a particularly strong interest in the issue of employment opportunities in broadcasting, issuing its own special set of equal employment opportunity regulations for broadcast stations, and by and large enforcing them.

Other reform issues failed to do as well. Most issues of content, such as excessive violence or sex and race stereotyping, were not dealt with by the FCC at all, in keeping with the Communications Act's stipulation that the FCC avoid direct involvement in content. The FCC also failed to enforce any standards of content on broadcasters, preferring instead to allow broadcasters to present almost any sort of proposal for programming they wished, with very little scrutiny, even as to the number or type of commercials aired.[1] Again, the age of activism focused public attention on these issues, and some groups were able to act on them in local petition to deny situations.

The broadcasting industry reacted negatively to criticism of content and by the end of the 1970s was asking Congress and the FCC for statutory relief from regulation that required attention to the needs and interests of the communities they served. Senator Barry Goldwater, a former broadcaster, became something of a spokesperson for the industry's cause, stating at one point his opinion that "a broadcaster's airwaves are *his* property, not the public's."

By the end of the 1970s the movement for reform and improved broadcast service found itself at something of a crossroads, stalled on its original path by new technological and political developments.

The movement's stall began in 1978 when the new chairman of the House of Representatives Communications Subcommittee, Rep. Lionel Van Deerlin (D-Cal.), introduced the first draft of his complete revision of the Communications Act of 1934. Van Deerlin had taken his new post the year before with the announcement that he planned a "floor-to-ceiling" revision of the old Act to take account of the emerging realities of telecommunications. The old Act did not even mention television, and certainly did not account for the development of cable and, as in the area of telephone regulation, could not deal with the emerging problem of the blurring of the lines between the new media.

Many people, both inside and outside the reform community, saw the Communications Act rewrite, as it came to be called, as a great opportunity and a great threat. Broadcasters undoubtedly felt the same way, as did phone companies, cable companies, and everyone else potentially affected by it.

The House subcommittee staff, under Harry (Chip) Shooshan, worked diligently on the problem of regulating communications in the public interest, and heard from a variety of sources. But when the proposed bill came out in 1978, it was, in the words of Dr. Everett Parker of the United Church of Christ's Office of Communication, "the greatest sellout of the public interest since the Teapot Dome scandal." Public interest and church groups saw in the proposed bill a total revision of the way communications is done, a revision along lines the broadcast industry would want. Very little appeared in the bill to benefit the public. Proponents of the bill argued that it included trade-offs, that although it totally removed license renewal, thus ending the local input into broadcast regulation, it gave a system of license fees in which broadcasters would pay the public for use of the airwaves. It became clear that broadcast opposition to the bill's license-fee provisions would result in a marked reduction in those fees, and public interest groups rallied in opposition to the bill.

A period of alienation resulted between the public interest groups and the subcommittee staff, who did not seem to understand how people who had been intimately involved in broadcasting in the public interest for years could be upset about what appeared to be a total sellout of the public interest. The staff position seemed to be that the goodies for the public in telecommunications were in the future technologies, not in the old "outdated" broadcasting which would be a lost cause by 1990 or 2000 anyway.

Broadcaster and citizen opposition to the original bill and subsequent bills in the House and Senate put the rewrite on the back burner through most of 1980 and into 1981, and it was then uncertain whether the new Congress, with its decidedly more proindustry and antiregulation perspective, would pick up the rewrite idea.

This experience had two important effects on the media reform movement. First, it gradually became clear that the reform movement needed to address the new technologies and their public benefits and harms because these technologies will rule the day long after broadcasting has gone. Second, the two-year process of fighting and working on rewrite legislation resulted in all other priorities being put on the back burner in favor of the immediate threat posed to broadcasting in the public interest.

Before the introduction of the rewrite, many groups had been working on a number of other issues. Much attention was being given to education and consciousness-raising on the opportunity of cable access channels and other cable services.

A coalition of church groups had initiated a new strategy on the problem of violence on television, taking shareholder resolutions to corporations in which they held stock, asking them to limit their involvement in programs containing excessive and gratuitous violence. This campaign achieved some success with a number of sponsors and was a factor, along with increased activism by the American Medical Association and the national Parent Teacher Association, in the reduction in the number of violent series on television in the 1978 and 1979 seasons. (The incidence of violence did not greatly decrease in the total schedule, but a large number of series which had relied on violence as a plot gimmick did go off the air.) In negotiations directly with the networks, this group was moving to the issue of sex and race stereotyping but suspended those activities to devote full attention to the rewrite.

Had the shareholder activism continued, its next likely strategy would have been to address the issue of prescreening of network programs by local stations. FCC regulations and the Communications Act hold that local stations are the sole arbiters of what goes on their airwaves. They cannot turn that responsibility over to anyone, not even to a network. All local network affiliates have clauses in their contracts with the networks that allow them to refuse to carry any network program they feel does not fit the needs of the local community.

One particularly activist broadcast company, Westinghouse Broadcasting, which owns five television stations in various cities, received quite a bit of press coverage for exercising that right in 1977. The incident occurred when ABC introduced "Soap," which had been advertised as promising to "break new frontiers in sexuality and frankness on television." Westinghouse felt that the network's salacious promotion of "Soap" and its refusal to allow stations to see "Soap" far enough in advance of its airing warranted their refusing to carry it on their Baltimore station.

Although the network cannot force local stations to carry its shows if the stations deem them inappropriate, it can pressure them to air shows by not allowing them to prescreen the shows in time to make alternate arrangements. When a local affiliate refuses a given show, it must be prepared to forfeit the money the network would have paid it for carrying the show, buy something else to air, and sell the advertising time at lower rates than the network shows could command. Thus, even though the option technically exists, stations have few incentives to actually refuse to clear (show) network programs, and real disincentives against doing so.

This issue could have been addressed in a strategy such as this: Shareholder actions could have been taken to networks to encourage them to regularly allow more lead time for local stations to prescreen, and to local stations to encourage them to develop standard procedures for prescreening. This would have given local affiliates more discretion over what is shown in each community. This strategy would not have been intended to set up anything like a local review board. Local broadcasters are also prohibited by the FCC from allowing such a board to decide what is aired. Rather, it would have encouraged broadcast licensees to do what they are licensed to do: serve as local trustees of the public interest.

These shareholder actions and other priorities took a back seat to legislative action in the late 1970s, and 1980 saw a threat to the public interest open on another front, at the FCC.

The Carter FCC had a decidedly different tone from earlier ones. Many people were encouraged when long-time liberal Senate staffer Charles Ferris was named chairman, seeing his appointment as a possible move toward more rigorous Commission enforcement of public interest standards in broadcasting. *Broadcasting* magazine, the industry's leading trade journal, was highly suspicious of Ferris, a positive sign to many in the religious and public interest groups. The public interest groups' hopes were also raised by Carter's appointment of Henry Geller to the top executive position in telecommunications, the director of the National Telecommunications and Information Administration. Geller was a former FCC general counsel and lawyer with the public interest Citizen's Communications Center law firm.

Most of their hopes were not realized. Geller began to voice his frustration with traditional ways of regulating broadcasting and his growing feeling that broadcasting is not regulatable. The career of this man, whose considerable abilities found him at the center of several very significant moves to strengthen public interest standards in broadcasting, reveals the transformation of opinion on these issues that overtook nearly the entire Carter communications regulation team, with the exception of Pertschuk at the FTC. Geller and the rest caught what came to be known as "Potomac Fever," the strange malady that spread through the Carter administration, blowing the winds of nearly mindless deregulation into every agency and bureau.

In communications, the message seemed to be that since the new age was dawning, ways of regulating in the public interest by deregulating needed to be found. The initiative in this area was taken by the FCC, which decided in February 1981 to deregulate radio entirely, doing away with nearly all mechanisms for public input into licensing, all limits on commercial practices, and all implicit and explicit guidelines regarding programming. It was widely felt that the move was only a precursor to later deregulation of television as well.

Thus the public interest movement was again drawn into a fray merely to maintain the status quo, rather than to advocate for the real changes that seem to be needed. Many reform movement representatives were frustrated by this turn of events. Don Matthews of the Telecommunications Consumer Coalition staff put it this way: "We end up sounding like we think the old FCC and the old way of doing things is the best . . . is what we want. Nothing could be further from the truth, but it looks like the lesser of two evils at this point."[2]

The focus of the FCC's argument, later picked up by proponents of deregulation on Capitol Hill, is that there is no such thing as "the public," that the public needs and interest can be assessed by what it is willing to pay for. No matter that there are those who cannot pay, those who will have no voice because no one will want to pay to hear them; the "market place" and "market forces" should rule. In fairness, it appears there is some well-meaning desire behind this thought. Geller said:

> I just don't see why we should be trying to make someone
> [broadcasters] do something that they really don't want to do . . .
> why not just get government out of the way, and let the people
> rule by voting with their pocketbooks.[3]

Their argument ignores the basic fact that in broadcasting, be it radio or television, the audience is not the customer. The free market situation envisioned by the FCC and others works when discussing a standard business-customer relationship in which customer demand can cause business to make fine adjustments in the goods and services it offers. Such demand also affects prices, making them go up when demand is high, and go down when it is low. That situation, however, does not describe broadcasting.

As was shown earlier, in broadcasting the audience is the *product,* not the *customer.* The customer is the advertiser, who already benefits from the most cost efficient audience-delivery service in the history of the world, American commercial broadcasting. To say that the audience can somehow "vote with its dollars" in television or radio is extremely naive. To their credit, many deregulators point out that they understand this. But, they say, public demand can still make itself felt in broadcasting by tuning in or out, thus affecting ratings, and thus affecting advertiser interest. While this may be true, the mechanism is still not directly responsive to audience demand; it operates only in interaction with the advertisers and their interest or lack of interest.

A still deeper and more profound reason exists for objecting to the idea of letting "market forces" dictate the public interest obligations of broadcasters. This is the abandonment of the "public" as an entity in favor of "consumers."

Les Brown, former TV columnist for *The New York Times* and now editor of *Channels* magazine, said it well in a 1981 address to the National Council of Churches' Communication Commission:

> This *laissez faire* policy [of deregulation] is really no policy at
> all. The regulators say, "Let the people decide—they'll 'vote' with
> their dollars." What is endangered here is the extinction of "the
> public" as an entity. The First Amendment is a responsibility of
> broadcasters, not their right, and defining your service in terms of
> "consumers" is qualitatively different than defining it in terms of
> "the public."
>
> The theory is that anyone who objects to what is on the air
> can turn to another service. . . . This is the way we've come to

regulate the most important social force in the land.

The fact of the matter is that these "channels" of choice are controlled by whoever has the monopoly. Under those circumstances 52 channels don't mean a variety of choices. . . . By doing this, we are taking control of these media away from the government and giving it to corporations. This results in going from monopoly to oligarchy. . . .

The activists have made broadcasting today attractive enough to be trusted by regulators. . . . . They have antagonized the regulators by opposing their giveaways to the industry . . . and so the doors to public participation are being shut.

Brown also addressed the process of the Carter FCC moving to deregulate broadcasting:

The "liberals" got over to where the conservatives had always been and that ended the debate . . . the other side wasn't present . . . they just said, "We just have to accept the fact that broadcasters are venal . . . we make them go through this long, expensive process [license renewal with public input] and they just go out and do what they want to anyway."[4]

This basic cynicism of recent well-meaning regulators was not too dangerous when it was held by those in power at the FCC and the NTIA. What is threatening in these developments is that the liberals' moves to deregulate will probably be followed by a more conservative Congress and FCC, resulting in a radical turn from the idea of public interest regulation during the 1980s.

These developments have been markers along the way to the crossroads at which the public interest media reform movement now stands. The optimism and opportunity of the early- and mid-1970s has been tempered now by realism, hard negotiations, and actions carried out. The public interest law firms that had served as a backbone of the movement, providing legal services to local groups as they sought action from their local stations, have been gutted by the withdrawal of foundation funding. The venerable National Citizens Committee for Broadcasting, the flagship of the reform movement for years, has been subsumed into the larger Ralph Nader organization. Although this has been good for the longevity of the NCCB and its important work, some took it as a sign of retrenchment. This has not proved true, however, with Nader's own Congresswatch organization beginning advocacy on communications issues.

Another important development late in the '70s was the formation of a new coalition to deal with the issues of the new age, the Telecommunication Consumer Coalition. The TCC was founded by three longtime activist groups, the United Church of Christ, the Consumer Federation of America, and Consumers Union. TCC acts as an umbrella and switchboard organization for regulatory and legislative issues in telecommunications.

TCC's first director was Dr. Ralph Jennings, a longtime associate director of the UCC's Office of Communication and widely respected public interest spokesman. He left the post in early 1981, and his successor was Janice Engsberg, also from the UCC. The development of the TCC, the continuation of the work of the NCCB under Sam Simon and former FCC Commissioner Nicholas Johnson, the Media Access Project, and the Citizens' Communications Center should all be seen as possible signs of realignment and growth. Unfortunately, the church groups that had been a key part of the movement for many years find themselves short on funds and some have discontinued their advocacy programs. But church and public interest groups' attention to these issues must and will continue.

The interest of churches in issues that might be construed as "secular" is one of the curious phenomena of the television and radio age. It was a church, the United Church of Christ, which in effect started the movement by pursuing the WLBT case. Media reform was seen at first as a logical extension of 1960s civil rights activism. But as the decade wore on, churches were increasingly interested in the media as institutions deserving social witness.

The United Methodist Church's Women's Division became concerned about the stereotypes of women on television, images that promoted ideas of women as second-class citizens or merely as sexual objects. Children's perceptions of life were being affected and this seemed as natural an arena of witness as seeking improved working conditions and opportunities for women.

The Church of the Brethren, United Church of Christ, United Methodist Church, Christian Church (Disciples of Christ), Presbyterian Church US, and the American Lutheran Church saw television violence as an important aspect of the total cultural fascination with and acceptance of violence as a method of problem solving. These groups were joined by many Catholic groups in this area of witness in shareholder resolutions.

The National Council of Churches and many local councils of churches and Catholic dioceses worked for improved opportunities for minority and female employment and visibility in broadcasting through attention to the Equal Employment Opportunity performance of local stations. In Cleveland, the Catholic Communications Office and Council of Churches continue a regular monitoring of EEO compliance, an activity which is a model for the rest of the country.

The NCC, the United Church of Christ, the US Catholic Conference, the Church of the Brethren, and many mainline and evangelical churches have used FCC proceedings and Congressional legislation regarding broadcasting as important forums for witnessing to the needs of the poor, disadvantaged, and dispossessed.

This work must continue. But at the threshold of a new age, there is the risk both of forgetting the past and of missing the best opportunities in the future. The reform movement must analyze carefully the developing technology and incisively witness to those aspects of it that are decidedly against their interests and the interests of the public at large.

The developments that have brought evangelical and political conservatives to power in recent years have also brought renewed national debate on the "quality" of television programming. Early in 1981, Ron Godwin of the Moral Majority, Eagle Forum director Phyllis Schlafly, persistent television boycott leader the Reverend Don Wildmon, and other conservatives announced plans to "clean up television" with a coalition called Coalition for Better Television.[5] CBTV may turn out not to be a censorship or vigilante group and it may do some good. But its prominence and its connection with religious conservatives may create difficulties for those who seek a better, more open television system expressing a wider range of views and opinions, rather than a narrower one.

When the churches were carrying out the shareholder campaign on television violence, they were often accused of being in favor of censorship. This was decidedly not the case. Their interest in reducing violence was based on a desire to see a different form of television service, one that relied on valuable experiences and images, entertaining or educational, rather than on negative plot "gimmicks" such as violence to attract an audience. The broadcasting industry is often too quick to holler "censorship" when a church or public interest group presents complaints about programming service and it is often too slow to try to understand the source of those complaints.

Broadcasters exercise censorship every day, every minute. That is precisely their business. Their whole activity is the selection of the images millions of Americans will see. The goal of the reform movement has been to assure that that selection is made not solely on the basis of what advertisers want or what the ratings say, but also on the basis of informed input from the public being served.

National legislation that prohibits showing certain things is probably not the answer. Neither are highly imperfect and imprecise campaigns using floods of letters, petitions, and boycotts that send the FCC and the broadcasters running to their bunkers yelling "censorship." One problem with these strategies is that they require little investment of the people who participate in them. Broadcasting is an important resource; it is one of the most valuable opportunities this society has to uplift, educate, and inform itself. Those who are interested in changing broadcasting and the face of new communication technologies should do so out of informed judgment, not out of fear or bigotry. They should be prepared to become media experts. They should begin to understand how the media work and what the people's rights are in regard to them. The media are not all powerful, but they can appear that way if church, minority, special interest, and public interest voices are not involved directly in speaking to them through participation in legislative and regulatory processes at the national and local levels.

So what should the media reform movement do now? A number of obvious directions for action should be pursued at the beginning of the new age of telematics. Concern for conventional broadcasting must continue. Most people will continue to get most of their media input from broadcasting well into the 1990s. (One optimistic projection for cable penetration suggested recently that by 1995 only 75 percent of all viewing would be of network television.) Attention to cable must continue. New ways of addressing other media developments must be found. Most important, renewed attention must be given to media education, the process by which viewers become informed and aware consumers of media. It will be helpful to look at each of these directions in some depth.

Conventional broadcasting will exist for some time. Certain aspects of it, such as the Fairness Doctrine (an FCC regulation requiring broadcasters to seek and present a variety of views on important issues—an opportunity for input from interested members of the public), will probably continue to be an obligation of broadcasters for some time. Efforts should be made to forestall as much as possible the trend to deregulate in directions that do away with the public trusteeship principle. Efforts also should continue to make trusteeship work on the local level. The prescreening campaign seems to be a natural issue. Local network affiliates should be encouraged to actively evaluate and plan the kind of programming they are showing. With more activism in this area, network offerings will change and could become more responsive to local needs and interests.

Children's programming and service to children in general should continue on the agenda. There is no reason why standards limiting the kind of advertising directed to children need to be set in Washington. They can be set locally, by local stations in consultation with parents, teachers, and local institutions. The same could be said for programming directed at children. Why could local activism not continue in this area? Greater cooperation with the community could prove to be in the interests of conventional broadcasters as they face competition from new services, services that might not be able to respond to the needs of local communities as well as local stations can.

Cable television should also stay on the agenda. Most cable systems franchised since the FCC dropped the regulations requiring public access have continued to offer access channels and services. They have done so to comply with requirements placed on them by local authorities. This trend will continue only as long as local communities and potential subscribers remain interested in such services. Once these services are in place, they must be used as much and as creatively as possible, or the cable system will be able to argue for their elimination because of lack of use.

Already, the cable industry seems to be preparing to try to eliminate local access as an option. Cable managers in systems with access and community service channels regularly complain that they are underused and underviewed. No one, however, has really demonstrated what underuse of access channels is. The channels may be valuable to the local community regardless of their level of use. Education and experimentation are needed to discover how access channels can be of service to the community and its institutions. Concerned persons need to become involved in the struggle to maintain cable community service.

On the national front, legislation already has been introduced once in Congress to make it illegal for communities to require any community service of cable franchisees. Such cynical and mercenary moves should be opposed by interested persons and groups or they will succeed for lack of opposition.

Many of the original 12-channel cable franchises (over 80 percent of the systems were still of this type in 1980) will come up for renewal in coming years. At that time, they are likely to try to have their franchises renewed with only minimal upgrading of service. Persons in communities where this will take place should begin to learn what the possibilities are so that renewed franchises can offer more channels of service, access and community channels, and facilities for local production of programs.

Cable is a lucrative business, even in small communities, and community service should be offered in exchange for such franchise rights. It is an exciting aspect that this process is entirely a local one. The opportunity exists for each community to evaluate and achieve the type of cable service it needs and wants without the intervention of outside authorities.

The trends in telecommunications seem to be toward centralizing control; eventually most cable and broadcast services will be subsumed into a few giant corporations. Already this trend is apparent with such corporations as The New York Times Company and Westinghouse Broadcasting (the fourth largest broadcasting company after the three networks) moving to buy cable systems in 1980 and 1981. Very few multiple system operators (MSOs) are likely to survive this trend. If the present regulatory situation continues, however, local communities will still grant franchises and will be able to direct the kind of cable services they want, in spite of the size of the corporation holding the franchise. The cable industry will seek to have the federal government outlaw such discretion by local communities, reasserting federal authority. Such actions should be opposed as they will forever seal the fate of cable, making it a provider of entertainment services and little else, save what the audience is willing to pay for.

Media reform efforts need to consider the coming plethora of other media, including the home video services discussed earlier. The issues of content, stewardship of time, impact on values, and interpersonal interactions that held in conventional broadcasting will hold in new video as well. Even more important, though further down the line, is the coming development of the "home information/communication center." The problem of cognitive overload discussed earlier and other problems not yet foreseen need to be addressed carefully.

Privacy is also an important issue. Public input and public interest debate must be involved in devising regulatory safeguards to insure that information, which can be gathered so easily in systems such as those of the age of telematics, is not used to the detriment of individuals or for their oppression.

The information poor are still another problem. One solution would be to have society mandate that basic access to information and information services be a right under the law, allowing subsidy if needed to give people access. Such a move would be entirely in the spirit of the U.S. Constitution, which provided for universal mail service (the most sophisticated communications technology of the 18th century) available to anyone regardless of where they lived, at the same cost, regardless of their ability to pay. Telephone regulation has followed the same pattern, with some states mandating the provision of "lifeline" phone service to those who need it at less cost than standard phone service. The people who need these services are by and large without voice unless others advocate for them. This would seem to be a natural area for witness, particularly for the church.

The strategies available to those interested in witness on these issues have changed in recent years, but for the present some redress of grievances will be available in the courts, and some regulatory machinery and legislative oversight over new technologies will continue. Beyond those avenues, the reform movement can use direct contact with the industries involved, including shareholder actions, meetings, and other activities. All of these openings must be used if the movement is to continue to have a dynamic relationship to new developments.

Especially important is attention to media education, helping media viewers and consumers to become informed and aware. The effect that regulatory action can have on specific programs consumed by viewers is shrinking. Therefore, people must be as aware and informed about the media environment in which they live as they are about their physical environment. What are its dangers? Its opportunities? What effects might it have? What are the best ways to use these new technologies?

This idea is not new. Many schools and churches have been doing this sort of media education for years. Others have achieved the same results by training people to produce their own media. The goal should be the creation of new skills and increased awareness, new persons for the new age.

One such curriculum, Television Awareness Training, was started by church groups in the mid-1970s to train critical viewers in a holistic way. The focus of T-A-T is on helping people learn to use television (and the new media) creatively, and to see the extensions of the media into other parts of their lives so that they are empowered to live decisively and autonomously in the media age. One of the greatest failures of broadcasting has been its failure to teach viewers about itself. It has utterly failed to help its consumers understand in any way the process in which they are involved as they watch. It is left to the churches, the schools, the homes to carry out that educational task. This media conscientization is a vital part of any program to improve the interaction between the media and people's lives.

Indeed, the reform movement is at a crossroads. It would be far easier to say, "Why can't we just hold off on all of this change until we've had a chance to decide what the effects of all of this will be?" But that is not possible. In this area, more than in almost any other, the constancy of change is the only constant. Developments are moving ahead and the reformers must climb on board or be left in the dust. Other societies have taken a more studied approach to these developments, and at some future time this society may do likewise. But for now, the telecommunications age will move ahead apace. The media reform movement, too, must move ahead with studied grace and power.

Chapter 12

The Challenge to the Concerned

A revolution is happening, a revolution that will change forever most of what we know and do. Those changes may be subtle ones, intentionally designed to be palatable, but they will also be powerful ones, touching nearly every aspect of life. This is not a time for passivity on the part of those interested in issues of equality, justice, and public service.

For the past 15 years many concerned people have been trying to make television and radio better. Their efforts have concentrated on service to children and other special communities, and on the possible effects of certain kinds of programming. Various strategies have been used. Some people have been content to write letters and make phone calls. Others have become involved in formal petitions to the Federal Communications Commission. Now all are facing possible disenfranchisement with the development of a plethora of new media and new services, opportunities that will change markedly the rules of the game.

Many claims are made for these new developments, including that they will democratize broadcasting by increasing consumer choice, that they will empower individuals through improved access to sophisticated services, that they will finally create that national or global "village" which has been so long touted and so long in coming.

These are myths, myths of a piece with those about television discussed earlier. Broadcasting and telecommunications services that emphasize consumer choice are no more possible in the new age than they were earlier unless concrete actions are taken to make them so. As Les Brown has said, just because there are 52 channels available does not mean there will be increased choice because all 52 channels still operate purely in the interest of the system that owns them and the advertisers who pay the bills.

Although the new age will put sophisticated services within the reach of the home, that will be the case only if those homes are willing or able to pay. The new age runs the risk of merely raising the financial threshold for entry into full participation in society. It will do nothing to improve the quality of that participation unless decisions are made to ensure that it does.

The creation of a global village in the future through these technologies is no more possible because of them. The ease of communication which will be possible will do nothing to change the will and desire of people to be so connected and involved.

In order to prepare for the future, it is important to observe and learn from the lessons of the past. Optimistic projections (however accurate or likely) must be evaluated in light of the present situation and what is known about how people and media have interacted in the past.

Most people will continue to use conventional television for some time to come. Thus, the issues of television content and its effects will continue to be important concerns. Because even the new age media will be the result of the industrial mass production process seen in television, the same tendency will continue for new media to carry dominant rather than unique values and ideals. The tendency for light entertainment appealing to the mass taste will continue to be the most profitable, and thus the most successful.

What has traditionally been known as news and public affairs programming will be changed in years to come as it has been in the past. In fact, the transformation of news into an electronic environment has been one of the most interesting phenomena of the electronic age.

Early radio and television news, typified by the contributions of people such as Edward R. Murrow, drew a great deal of its moment and force from the traditions of print journalism. Television news became an institution in its own right more recently and has been credited with transforming the nation's relationship to important issues and experiences, such as the Kennedy assassination and later the Vietnam War and the Watergate scandal.

Just as broadcast news could borrow the traditions and credibility of print journalism, it could not escape some of that medium's tendency to sensationalism and commercialism. For many years, the power and impact of journalistic tradition carried the day in electronic journalism.

But in the early 1970s, local station managers began to realize that local news was one of the most important profit centers of local operations, and began to look for ways to increase ratings so that more local and national advertising dollars would flow their direction. The phenomenon of the news consultant emerged, a specialist who could tell local stations exactly how to dress up their local news to beat the opposition. Many of the now-familiar artifacts of local news style that emerged late in the decade are the direct results of such news consultations. They involve shorter stories, more stories, stories that emphasize film (often overlooking stories with no visual images), the "youth image," feature stories instead of news stories, and, of course, "happy talk," where much time is taken with meaningless banter among the newspeople intended only to create a friendly image.

Marilyn Baker, a television reporter in San Francisco, has said that the formula on which local news was created was a pat list of stories that must be carried every night, regardless of their interest or importance. These included, according to Baker, a fire, a sex crime, a murder, a robbery, and a "kid-dog" story (meaning a light "good news" item of some kind).[1] Local news gradually began to see itself primarily as entertainment, bound by the demand that it pull its weight in the ratings race.

These trends on the local level gradually moved to the national level during the '70s. "Happy talk" transformed all electronic news, nowhere more vividly than in NBC's venerable "Today Show." Locked in a ratings race with the upstart ABC "Good Morning, America," "Today" gradually adopted that show's style, selecting more photogenic personnel to update the rather staid image it had projected earlier. "Today" even added a goofy weatherman (a staple of local news) and, in 1981, began to carry as a regular feature the Hollywood gossip reports of Rona Barrett.

This is not to say that this has all been bad, or that no news is now done on programs such as "Today." More discriminating viewers had the option during this time of the "CBS Morning News," which concentrated much more on hard news and took a beating in the ratings. People apparently like their news to be entertainment as well.

The new age of telecommunications will bring changes in many areas. There will be more diverse sources of news, perhaps some that will offer alternatives to the limited or dressed-up style that has come to dominate broadcast fare.

But those who wish for a different sort of broadcast news

coverage may or may not be able to get what they want from the new media. Even if there is a large enough market to justify such alternatives, there is nothing to guarantee that they will develop if other services such as old movies and sports seem to be more lucrative. (It is becoming axiomatic that the American public's appetite for entertainment and sports is boundless.) Broadcast news developed in the first place not because there seemed to be a market for it but because the Federal Communications Commission and Congress expected that these public trustees they were licensing should offer news as part of their service to their communities. Without that expectation, it is unclear whether even so limited a service as now exists would have developed on its own.

The problem with leaving the development of such services to the market is that the market is already heavily conditioned in its expectations for news by what it has already known. No one can really say what other options for news services are possible or might have developed. The structure of American broadcasting under the Communications Act, which granted full-time, noncompetitive licenses to broadcasters nearly in perpetuity, has saddled the U.S. with a system remarkably nondiverse and nonexperimental. It is at least an even bet that the future "diverse" technologies will not be an age of great choices and options, but rather will be an age of programming services that build heavily on the past, still seeking to maximize audience instead of offering alternatives.

There is good reason to expect this to occur. The Public Broadcasting Service was to have been an alternative where commercial pressures would be abolished, minimizing the need for programming to compete for audience, thus providing more diversity. Generally, this has proved to be true. Public station schedules are more diverse than those of commercial stations, but they also reflect a tendency to provide programming to elite interests, not "something for everyone." Still, one of the major concerns of PBS stations must be the size of the audience they gather. They cannot expect to continue to receive corporate grants, public subsidies, or even subscription support if their audience is small and noninfluential.

This trend toward maximizing the audience must be ended as the new age dawns. There may be opportunities and options not yet discovered. The current trend to see the audience size and its dollar power as the only determinants of what services will be tried or made available may tie the media to old paths and old expectations just at the time when new opportunities might be available.

For viewers, that means being open to new things. Much of viewers' reluctance to really exploit their opportunities to influence television and cable is because they have grown accustomed to things the way they have always been. Looking at a new program or service, they turn away because it is not as well-produced as commercial broadcasting. A local program is turned off because it is not as sophisticated as a network offering. A cable access program is pronounced awful. Viewers must acknowledge that change has to start somewhere; they must be open to media that can be free to bring some absolutely horrible things, but also to bring some great things. Viewers should be prepared to be critically aware of whether what is being offered is actually better and more diverse, or only superficially so.

Beyond the likelihood that future programming will not differ markedly from that of the past, another long-term trend bears consideration. That is the tendency already seen towards concentrating power in the media in progressively fewer major corporations. Already, many small cable companies have been purchased by other, larger media organizations, including *The New York Times*. The pay cable company owned by Rockefeller Center shocked the Public Broadcasting audience in 1980 by moving in ahead of PBS for the rights to BBC programming, long a staple of PBS fare. (Public Broadcasting announced plans for its own subscription channel that same year.) Other corporate mergers and acquisitions are expected, with some observers projecting that the ultimate result will be a handful of media conglomerates controlling publishing, television, cable, and new technologies. (The tendency for electronic media companies to acquire print interests and vice versa became quite pronounced in the 1970s.) One distinguished study of the future projects that rather than a handful of companies in this area, one, AT&T, will be the ultimate provider of nearly all services. In this era of change and concentration, new legislation may eventually prove necessary to separate control of content from provision of the network. Some voices are already calling for such change.

The effects of this media concentration will reveal themselves to ordinary users in what they have come to expect from broadcast services. One such change is the move of BBC to a pay cable channel. Another change will be in sports programming. It is likely that the 1988 Olympic Games will not be available on television except to those who wish to pay to see them over pay services. Already, one major league baseball team, the Chicago White Sox, has announced plans to move from conventional broadcasting to cable because pay cable offered a more attractive package. The Super Bowl will not be on conventional television for long, nor will many other professional sports events. The winds of change are in the air.

What are the strategy options open to interested people as the telecommunications revolution begins? The traditional methods of input to conventional broadcasting still exist. Interested people can file comments and petitions in FCC proceedings regarding their local stations. Expressing concerns directly to stations and networks can still have some effect. It is better to write than to call, and good practice to send copies of correspondence with local stations to the FCC for their files there. Such things are reviewed at license renewal time. Stations also keep files of letters, comments, and their applications and programming promises (made as part of the license applications to the FCC) on file for public inspection during regular business hours. People who have an interest in a particular station can look through that file.

With regard to cable, the primary avenue of input is through the local community and local government. Cable companies have fewer standardized obligations than do broadcasters, but they have responsibilities nonetheless. Particular interest should be taken in those situations where cable systems are applying for new franchises, upgrading current ones, or otherwise appealing to local governments. People who have an interest in the issues addressed here should become involved in those processes.

Congress and the FCC will continue to make decisions regarding the development of the various technologies of the new age for years to come. Christians, like other citizens, should become knowledgeable and informed about these developments because many of them will have lasting impact on life now and the lives of coming generations. The media reform organizations in Washington, mentioned earlier, can provide information about the developments there. They deserve the support and involvement of concerned people.

Above all, concerned people should become active, informed, critical consumers of these new technologies and services. The finding that purchasers of new technologies actually spend more time than ever with their television sets should serve as a lesson. Far from bringing a time when increased choice and diversity makes better consumers of everyone, these technologies merely come on the scene and present themselves. It is up to the consumers to exercise stewardship over them.

The fact that a program on cable or satellite (being provided by Public Broadcasting, the Christian Broadcasting Network, or even CBS) must be paid for does not automatically mean it will be a high-quality program. Television and other media must be used responsibly and children must be taught to do likewise. Subscribing to a "children's channel" on cable may not really be more responsible parenting than using Saturday morning cartoons as a babysitter if what children desperately need is more interpersonal contact with their parents and peers.

The content of programming will also bear scrutiny. Some changes in the evenness of quality of the programming will be noticeable. The new media will not subscribe to the broadcasters' code (administered by the National Association of Broadcasters, not the FCC) so things may be shown on the new media that have not been seen on conventional television. If this occurs, and a greater diversity of programming is offered, viewers will have a special responsibility to cast off the easy complacency of watching television in the old days without running the risk of seeing anything really new or different. Viewers' skills at seeing and evaluating the values messages in the media they consume will need special attention.

Afterword

This book is the work of someone who really likes television, a child of the television age. I can remember life before TV, but ever since my family first got it, I have watched a great deal of it. I feel a great loyalty to television in many ways. It brought me images and ideas I could not have had otherwise. It took me places I could never have seen on my own, and taught me much about the world.

As I have studied television, I have gradually evolved from an unabashed lover of TV to someone who is perhaps more pragmatic and skeptical about the medium, certainly not one hostile to it. In fact, it is the rich potential of the medium that drives me and others to seek changes in the way it is managed by its institutions and used by its viewers. I am not an expert on what television should be, nor how people should use it. Rather, I urge all of us to work for ways that it can be as responsible as possible to all the needs for its services that exist. The new telecommunications age could be an opportunity for that to happen, or it could be more of the same, with the same institutions in charge, the same sorts of programming available (only more of them), and basically the same interests served, those of the advertisers. Business enterprise should be involved in the provision of telecommunications services, but it should serve at the pleasure of the people who need some of the things these technologies have to offer.

My interest in specific areas of television rises out of my background and upbringing, which stress faithful commitment to Christian principles of honesty, fairness, justice, humility, simplicity, and service. Because of my commitment to pacifism, I was very early concerned about the militaristic and violent content of television programming and news. My interest in fairness led me to advocate for access to television airwaves and telecommunications services on behalf of the poor, the disadvantaged, the undereducated. My concern for humility, simplicity, and service drew me to an interest in the commercialism of broadcasting, its stress on salvation through belongings, and its glorification of individuality at the expense of community. Television does not deal as regularly in my basic values as it does in their opposites. I do not assume that my values should prevail on the air, but I can work with other people who also see their interests left out of this important medium to see that it does a better job of serving all.

When I search for a theological grounding for my concerns about television, I come back to the profound lesson given me by one of my teachers, Dr. Davie Napier. Napier is an Old Testament scholar who has devoted much thought and reflection to the way that lessons from the Old Testament prophets can inform contemporary life.

His search led at one time to the account in 1 Kings of the confrontation between Elijah and the prophets of Baal at Carmel. Napier saw in that confrontation the model for our contemporary confrontation with a society and culture that presents us with its owns prophets, its own law, its own practices, and its own allegiances. The message of Elijah at Carmel should be a clear call to each of us at this time. Napier puts it this way:

> Baal worship takes place whenever and wherever that faith becomes provincialized, parochialized, and accommodated to the culture in such a way that the adherents lose altogether the sense of critical distinction between Yahweh and Baal, between the Word of God and the word of persons, between the word of earth and the word of the system, between God who is and god who is made, God who creates and god who is created—in sum, between God and his cultural image or, more bluntly, between Christ and Mammon. Jesus said, "You cannot serve both," knowing full well that this was precisely the prevailing religious situation of his own people.[1]

One way of seeing the theological challenge of television (indeed, these new media, be they cable, video games, or computer publishing) is to realize that it is through these technologies that a major portion of our society's claim on us and on our time and commitments takes place. I am not saying television is an instrument of the devil. Neither am I saying that these technologies are somehow theologically or ethically neutral. They are, by virtue of their particular place in the social, economic, and intellectual life of our culture the means by which the demands of earth are made to press on us. As such, they take on a theological dimension which involves our being careful, circumspect, and temperate in our use of them. At times we should stand, with Elijah, and call to accountability these institutions that have come to stand so much for mercenary, vainglorious secularism in our society.

My concern about television is not universal, even among those groups who would seem such logical allies. A certain antitelevision bias has always existed in the United States among both secular and sectarian intellectuals. Television and the new media seem trivial in the face of the major crises of the day. For many people, someone interested in studying television dredges up memories of the "media freaks" they knew when they were younger, the kids who wanted to run the projector when a film was shown at school, the kids who became the ham radio operators, the CB operators, the owners of home computers.

The time has passed when we can afford the luxury of trivializing the developing electronic media. They are gradually but inevitably changing our lives. They already convey much of what we know about the very world crises that seem more pressing. The media are now at a state of development that requires well-informed people to add to their repertoire of skills the ability and knowledge necessary to be critically aware of telecommunications media and the various ways they affect our lives. Media awareness is no longer a luxury, an affectation, or a hobby. It is now part of adult basic education. In the future, no one who wishes to develop expertise in the disciplines of teaching, ministry, counseling, or even parenting will be able to consider themselves prepared unless they have also dealt with the development of basic media awareness, consumption, and advocacy skills. Neville Jayaweera, a long-time student of these matters and now a prominent spokesperson for Christian involvement in media advocacy, says:

> We are rapidly entering a new age . . . An age which has been called "the information age." This new age will be characterized by the industrialization of the communication process in the west. . . . It will be a time when the study of communication will cease to be simply "a discipline" and become instead the primary discipline by which we know and assess the activity of societies and cultures.[2]

I like television. I think I will like the media of the new age also. I look forward to having a home computer with which I can compose documents such as this more efficiently, eventually publishing them by computer as well. I look forward to having much of the gadgetry. But I also look forward to the challenge of continuing what I have tried to do in the age of television . . . to make myself a dynamic, critically aware consumer and advocate with regard to the system.

We must be ready to become "new persons for the new age." The times call us to conscientization and activism. We must learn, act, and grow as though the survival of our way of life depended on it, because it may. The telecommunications age has the power to transform us and everything we know. Unless we are to be its victims, we must take the initiative.

Stewart M. Hoover

NOTES

Preface

1. Elizabeth Eisenstein, *The Printing Press as an Agent of Change* (New York: Oxford University Press, 1979).

Chapter 1

1. Barry Cole and Mal Oettinger, *Reluctant Regulators* (New York: Addison-Wesley, 1978).
2. Robert S. Alley, *Television: Ethics for Hire?* (Nashville: Abingdon Press, 1977), p. 20.

Chapter 2

1. Charles Wright, *Mass Communication: A Sociological Perspective* (New York: Random House, 1975).
2. Ibid., p. 8.
3. "Canalization" is the process whereby mass media do not stimulate thought directly along certain lines, but instead channel persons' opinions in broad directions through a variety of means and functions. The concept comes from Paul Lazarsfeld and Robert Merton, "Mass Communication, Popular Taste, and Organized Social Action" in Wilbur Schramm and Donald Roberts, eds., *The Process and Effects of Mass Communication* (Urbana, Ill.: University of Illinois Press, 1974), pp. 574-575.
4. George Gerbner, "Teacher Image in Mass Culture: Symbolic Functions of the 'Hidden Curriculum,'" in Gerbner, Gross and Melody, eds., *Communication Technology and Social Policy* (New York: Wiley, 1973).
5. R. M. Liebert, J. M. Beale and E. Davidson, *The Early Window: Effects of Television on Children and Youth* (New York: Pergamon Press, 1973), p. 1.

6. Ibid., pp. 41f.
7. R.K. Baker and S.J. Ball, eds., *Violence and the American Media* (Washington: U.S. Government Printing Office, 1969), pp. 593-614.
8. Hearings before the United States Senate Subcommittee on Communications of the Committee on Commerce, March 1972, p. 28.
9. Stewart Hoover, "A Cognitive-Developmental Study of Moral Development and Television." Unpublished master's thesis. Berkeley: Pacific School of Religion, 1976.
10. There is presently great controversy over the utility of Kohlberg's theory. His formulation is, however, a useful tool to analyze reasoning systematically.
11. *Los Angeles Times,* September 9, 1978.

Section II, Introduction

1. Indeed, *Broadcasting* magazine reported (June 1980) that one of the first televised critics, Ron Hendron, was replaced that year by NBC to the great satisfaction of local stations who thought his reviews of NBC programs were too often negative. His replacement may be better, and CBS' counterpart, television scholar Jeff Greenfield, may be of a different kind. But even comprehensive reviews of upcoming shows would still leave the audience largely uninformed about the basics of television beyond the content of programs.
2. William F. Fore, *Image and Impact: How Man Comes Through in the Mass Media* (New York: Friendship Press, 1970).

Chapter 3

1. William F. Fore, "The Price of Television." Unpublished report, 1972.
2. Proctor & Gamble marketed the following products on TV during the 1970s: Big Top Peanut Butter, Biz, Bold, Bonus, Bounty Towels, Camay, Cascade, Charmin Paper Products, Cheer, Cinch, Clorox, Comet, Crisco, Dash, Downy, Duncan Hines, Duz, Folger's, Gain, Gleem, Head & Shoulders, Ivory, Jif, Joy, Lava, Mr. Clean, Oxydol, Pampers, Prell, Puff, Safeguard, Scope, Secret, Spic & Span, Tide, Top Job, Zest. (List derived from "The Television Sponsors Cross-Reference Directory."

Everglades, Fla.: Everglades Publishing Co.)
3. From "A Short Course in Cable," *Broadcasting-Cable Year-book* (Washington, D.C.: Broadcasting Publications, Inc., 1980).
4. The story of how he avoided the scrap pile by doing his own off-network syndication is a fascinating one. In this regard, Welk and "Hee Haw" blazed a trail in competition with the networks which made them pioneers of the new television age. Welk is now on more stations on his own than carried him in the old network days.

Chapter 4

1. George Gerbner, "Cultural Indicators: The Third Voice," in Gerbner et al., *Communication Technology and Social Policy,* p. 568.
2. Gerbner, Gross, Morgan, and Signorielli, "The Mainstreaming of America: Violence Profile No. 11," *Journal of Communication.* Vol. 30, No. 3 (Summer 1980), p. 24.
3. Ibid., p. 14.
4. Ibid., p. 15.
5. Tony Schwartz, *The Responsive Chord* (New York: Anchor Press/Doubleday, 1973), p. 52.
6. *Broadcasting,* May 5, 1980, p. 30.
7. Ibid., p. 31.

Chapter 5

1. Erik Barnouw, *The Sponsor: Notes on a Modern Potentate* (New York: Oxford University Press, 1978), pp. 85-86.
2. *Broadcasting,* September 8, 1980.
3. Hermenio Traviesas. Speech to the Chicago Chapter, National Academy of Television Arts and Sciences, October 16, 1976.
4. This "sensing of the American mood" on the part of the net-works is a fascinating area, one deserving much further probing than is possible here. My suspicion has always been that this "sensing" is a *protectionist* rather than an "active seeking out" process. I also suspect that the "mood" sensed by network sensors is the mood of their particular social group, tempered by in-ferential data from outside. They have often claimed that it derived largely from mail and other feedback, but Traviesas

went on to say that "organized protest by mail" is discounted. James Brown, a former consultant to CBS, reported the same thing at a conference of religious broadcasters in 1978. (James Brown. Speech to the North American Broadcast Section/World Association for Christian Communication, Phoenix, Arizona, December 2, 1977.)

5. *New York Times,* January 12, 1978.
6. Esther Rolle. Personal interview with the author.
7. John P. Murray. Quoted in preface to Liebert et al., *Early Window.*
8. George Conklin in "Wholly Mackeral Productions."
9. Jeff Greenfield, "TV is Not the World," *Columbia Journalism Review,* May 1979, pp. 29-34.

Chapter 6

1. Sec. 326, Communications Act of 1934.
2. Their refusal to do so has stretched to a benign attitude toward a whole class of stations—religious ones—which may have resulted in a violation of the First Amendment in reverse. A very good analysis of this situation by attorney Linda Lacey suggests that by exercising benign neglect, the Commission has encouraged religion.
3. There is great controversy over what constitutes subliminal techniques and how often such techniques are actually used. Blatant subliminality, such as the implanting on the television screen of suggestive words that flash by too fast to be consciously perceived but that are nonetheless "seen" unconsciously, is illegal. More subtle forms of subliminal suggestion may be present in all forms of advertising. A particularly well known treatise on this subject is Wilson Brian Key's *Subliminal Seduction* (New York: Signet, 1974).
4. Cole and Oettinger, *Reluctant Regulators.*
5. Since not all stations have subscribed to the NAB code, there are stations that ignore some of its structures. An independent station in San Jose, Càlifornia, for instance, began carrying contraceptive advertisements in 1975.
6. William F. Fore. Personal Interview with the author Feb. 1981.
7. Cole and Oettinger, *Reluctant Regulators,* p. 38.
8. Ibid., p. 43.
9. The stations that have lost their licenses for these reasons are:

WSWG AM and FM, Greenwood, Mississippi, and the Alabama Educational Television Network, which lost its five television licenses in 1978.

10. *West Michigan Telecasters v. FCC,* 47 USC 309 (d) (1).
11. Southern Regional Council, Atlanta, Georgia. Steven Suitts. Personal interview with the author.
12. Howard Symons, "Making Yourself Heard (and Seen): The Citizen's Role in Communications," in *Telecommunications Policy and the Citizen,* Timothy Haight, ed. (New York: Praeger Scientific, Inc., 1979), p. 17.
13. Cole and Oettinger, *Reluctant Regulators.*

Chapter 7

1. Jeremy Tunstall, *The Media Are American: Anglo-American Media in the World* (New York: Columbia University Press, 1977), p. 263.
2. Roy Nehall. Presentation to the Annual Assembly of the North American Broadcast Section/World Association for Christian Communication, Fort Lauderdale, Florida, November 30, 1974.
3. Tunstall, *Media Are American,* p. 39-40.
4. Herbert I. Schiller, "Mass Communications and American Empire," in *The TV Establishment: Programming for Power and Profit,* Gaye Tuchman, ed. (Englewood Cliffs, N.J.: Prentice-Hall, 1974), p. 177.
5. Cees Hamelink, *The Corporate Village* (Rome: IDOC International, 1977), p. 11.
6. Dean MacBride Commission, *Many Voices, One World* (Parish: UNESCO Press, 1980), p. 37.
7. This phenomenon is underscored by the existence of a cadre of people who have a hobby of collecting these "bus plunge" items from *The Times.*
8. An interesting sidelight in Epstein's study was his finding that this "hegemony and imperialism" by news crew deployment functions in the United States as well, with the major Eastern cities receiving the majority of coverage at the expense of rural areas in the West and South.
9. A number of major corporations (Nestlé, American Home Products, Bristol-Meyers, and others) were identified as having actively promoted infant formula products in poor areas of

developing countries, in spite of the fact that they were not widely needed and could actually prove harmful in some cases.
10. MacBride Commission, *Many Voices, One World*, pp. 37-38.

Chapter 8

1. The reasons why such activity is not illegal dates to a series of landmark Supreme Court decisions regarding cable transmission of a radio broadcast in public places, such as bars and hotels. In an early case, *Buck v. Jewell-LaSalle Realty Co.* 283 U.S. 191 (1931), the Court held that such carrying of a radio signal by a hotel for the enjoyment of patrons was an infringement of the copyright act. That doctrine held until the 1968 Court decision in *Fortnighly Corp. v. United Artists Television* 392 U.S. 390 (1968) when the Supreme Court overturned a lower court ruling that *Jewell-LaSalle* applied to cable television as well. Acknowledging that cable is a rather complex form of receiver, the court nonetheless found that copyright applies only when a signal is transmitted, not when it is received. Thus, the Court found that cable TV is a "signal enhancement service" not an origination one. This ruling was broadened still further by the Court in its later *Teleprompter* decision. *(Teleprompter Corp. v. CBS* 415 U.S. 394, 1974.)

2. Historian Eric Barnouw reports that an immediate result of this was a decline in local television production, where much material was now available to stations more cheaply through arrangements to use already made films. Erik Barnouw, *The Tube of Plenty* (New York: Oxford University Press, 1975).

3. They were upheld in this case which went to the Federal Court as *U.S. v. Southwestern Cable Company*. The commission had decided that under the doctrine in the Communications Act which allowed them to regulate matters "reasonably ancillary" to broadcasting, they could step in to cable. They chose to protect a local broadcaster in San Diego by restricting the number of distant signals the local cable company could import. *U.S. v. Southwestern Cable Co.* 392 U.S. 157 (1968).

4. Susan Greene, "The Cable Television Provisions of the Revised Copyright Act," *Catholic University Law Review*, 27:263 (1978), p. 278.

5. Home Box Office is owned by Time, Inc., publishers of *Time* and *Life* magazines.

6. Federal Communication Commission figures in *Broadcasting-Cable Yearbook*.
7. Jeremy Weir Alderson, *"Everyman TV,"* Columbia Journalism Review, January/February 1981, p. 42.
8. John Wicklein, "Wired City, U.S.A.," *Atlantic Monthly*, February 1979, p. 37.
9. *Broadcasting* magazine, May 1980.
10. Electronic funds transfer (EFT) is the electronic movement of cash between bank accounts without the necessity of paper drafts.
11. Wicklein, "Wired City," p. 40.
12. *Philadelphia Inquirer*, December 30, 1980.
13. Ibid., December 1, 1980.
14. John Bloom, "Invasion of the Cable Snatchers," *Texas Monthly*, March 1980, p. 93.

Chapter 9

1. This is probably a good thing. In spite of the decree, Bell has found itself the target of hundreds of suits in the years since, many involving claims by other companies of monopolistic and anticompetitive practices. Bell's most famous recent activity in this area had been its refusal, until forced to do so, to allow any non-Bell telephones or equipment to be used by subscribers. In order to keep up its defenses in these areas, Bell is very active in legal proceedings. The Associated Press carried an item in 1980 from the *National Law Journal* which reported that AT&T is the largest single employer of attorneys in the country, employing 863 of them full-time. The largest private law firms in the U.S. employed only 512 and 288 respectively. The FCC legal staff numbers around 70.
2. Arthur D. Little, Inc., *Telecommunications and Society, 1976-1991: Report to the Office of Telecommunications Policy, Executive Office of the President* (Springfield, Va.: National Technical Information Service, 1976).
3. Networks also use satellites but only for feeding signals such as sports coverage or news stories back to New York where they go into the telephone network distribution system and are carried thus to local stations. Public television made news in 1975 when it began feeding its programming to stations over satellite. It continues to be the only network to do so.

4. This is legal because they do not *own* their signals. The courts have held (as discussed above) that it is not illegal to derive income from receiving or enhancing reception of signals. If it were, it would be illegal for Sears to sell and install television antennas, for instance.
5. Both Turner and WPIX are, of course, building on early alternative news done by Public Broadcasting, where such programs as the "MacNeil/Lehrer Report" and "Washington Week in Review" offer additional TV voices to the commercial networks. WPIX's experiment is doubly interesting because a large part of its initial advertising income came from the Mobil Oil company after it had experienced a "falling out" with conventional commercial network news. Mobil had wanted to buy air time on "CBS News" in particular to rebut what it saw as the unbalanced coverage there of oil company profits; CBS consistently refused.
6. It should be noted that this also means that those households with VCRs actually watch *more* television than those without. Rather than becoming selective, critical viewers, VCR owners spend more time than ever with TV.
7. *On Computing,* February 1981.
8. *The Information Age,* an Aspen Institute film produced by Marc U. Porat, 1980.
9. Ibid.

Chapter 10

1. *Newsweek,* October 25, 1976.
2. Martin Marty, "The Invisible Religion," *Presbyterian Survey,* May 1979, p. 13.
3. William F. Fore, "The Electronic Church," *Ministry,* January 1979, p. 5.
4. Valuable histories of religious broadcasting are J. Harold Ellens, *Models of Religious Broadcasting* (New York: Eerdmans, 1974), concentrating on the nonevangelical churches, and Ben Armstrong, *The Electric Church* (Nashville: Thomas Nelson, 1979), covering the evangelical and new "electronic church" groups.
5. The NCC could not have prevented member groups from purchasing time anyway if they had wanted to; its policies are advisory only.

6. Jeffry Hadden, "Some Sociological Reflections on the Electronic Church." Paper presented at the Electronic Church Consultation, New York University, February 6-7, 1980.

7. It is my suspicion that most mainline churches will first enter the new age not with nonprint media, but only when technology catches up with their print activities, making computer-publishing and computerized distribution possible. As 1981 began, there were moves by some NCC-related groups to use satellite time for program distribution on a very limited basis, but this will probably not become a major enterprise anytime soon.

8. Linda Jo Lacey, "The Electric Church: An FCC-Established Institution?" *Federal Communications Law Journal* 31 (2) 1978, pp. 235-275.

9. Another similar rumor started over the showing of "a film about Jesus' sex life" on American TV in 1978. No such film was in the works at the time as the producer was unable to make it and it died as a project—but as a rumor it lives on. An unsuspecting newsletter called *The Modern People News* was inundated with mail and petitions when it inadvertently carried a story about the ill-fated film in 1980.

10. Armstrong, *The Electric Church*, p. 154.

11. Charles Swann. Presentation to the Electronic Church Consultation.

12. Pat Robertson. Presentation to the Electronic Church Consultation.

13. Robert M. Liebert. Presentation to the Electronic Church Consultation.

14. Edwin Diamond, "God's Television," *American Film* 5 (5), March 1980, p. 30.

15. Hadden, paper at Electronic Church Consultation, p. 10-11.

16. Mary Murphy, "The Next Billy Graham," *Esquire*, October 10, 1978, p. 29.

17. Ibid., pp. 30-31.

18. George Conklin. Presentation to the Board of Managers, Communication Commission, National Council of Churches, September 1980.

Chapter 11

1. There are guidelines, but they were established only to sanction industry practice, outlined by the industry's own NAB Code.

For a thorough review of these issues, see Cole and Oettinger, *Reluctant Regulators.*

2. Donald Matthews, S. J. Personal interview with the author.
3. Henry Geller. Speech to the Communication Commission, National Council of Churches, February 1979.
4. Les Brown. Speech to the Communication Commission, National Council of Churches, February 1981.
5. *Broadcasting* magazine, February 9, 1981, p. 27.

Chapter 12

1. Marilyn Baker. Speech to Women in Media Conference, Pacific School of Religion, Media Center, April 1974.

Afterword

1. Davie Napier, *Word of God, Word of Earth* (Philadelphia: Pilgrim Press, 1976), p. 46.
2. Neville Jayaweera. Personal interview with the author.

INDEX